Language Constructs for Describing Features

Springer-Verlag London Ltd.

Stephen Gilmore and Mark Ryan (Eds)

Language Constructs for Describing Features

Proceedings of the FIREworks workshop

Springer

Stephen Gilmore, BSc (Hons), PhD
Laboratory for Foundations of Computer Science, The University of Edinburgh,
Edinburgh, EH9 3JZ

Mark Ryan, BA, MA, PhD
School of Computer Science, University of Birmingham, Edgbaston, B15 2TT

ISBN 978-1-85233-392-8   ISBN 978-1-4471-0287-8 (eBook)
DOI 10.1007/978-1-4471-0287-8

British Library Cataloguing in Publication Data
Language constructs for describing features : proceedings
  of the FIREworks workshop
  1.Application software - Development - Congresses
  2.Telecommunication systems - Congresses 3.Communications
  software - Congresses 4.Banks and banking - Data processing
  - Computer programs - Congresses
  I.Gilmore, Stephen II.Ryan, Mark, 1962-
  005.7'1265

Library of Congress Cataloging-in-Publication Data
A catalog record for this book is available from the Library of Congress

Typesetting: Camera-ready by editors
Printed and bound by the Athenæum Press Ltd., Gateshead, Tyne & Wear
34/3830-543210 Printed on acid-free paper SPIN 10784614

Contents

VI

Preface

The concept of *feature* has emerged in the telecommunications industry as a way of describing optional services to which telephone users may subscribe. Features offered by telephone companies include Call Forwarding, Automatic Call Back, and Voice Mail. Features are not restricted to telephone systems, however. Any part or aspect of a specification which the user perceives as having a self-contained functional role is a feature. For example, a lift or elevator system may exhibit such features as: parking when idle; landing-call preference when more than two-thirds full; and compartment-call cancellation when empty. Thinking in terms of features is important to the user, who often understands a complex system as a basic system plus a number of features. It is also an increasingly common way of designing products.

This volume reports the proceedings of a workshop held on *language constructs* which have been deployed to describe features, in May 2000 in Glasgow. Themes of the workshop included:

- Syntax and semantics of feature constructs,
- Non-monotonicity of feature integration,
- Methodologies for feature-based specification and design,
- Feature-oriented architectures.

By focussing on language constructs, the Workshop is intended to complement the *Feature Interactions in Telecommunications and Software Systems* series of workshops which have taken place since 1992. Our workshop was colocated and immediately prior to the sixth instance of that series.

The workshop was organised as part of the FIREworks ESPRIT Working Group funded by the European Union (1997–2000). The aim of FIREworks was to provide a feature-oriented approach to software design. In particular, this encompasses the development of:

- Several feature-oriented specification languages (based on several existing languages).
- Verification methods for detecting and explaining conflicts between features; deducing consequences of feature combinations; and methods for constructing test suites for feature combinations.
- Several case studies assessing the adequacy of the languages and tools.

In addition to talks from members of the group in this volume, we were glad to welcome many contributions from other research workers in the field. Our thanks go to all of the authors for their timely submissions of the papers, and to the reviewers who helped us select among the submissions.

Stephen Gilmore   Mark D. Ryan
The University of Edinburgh   The University of Birmingham

List of participants

Marc Aiguier ⟨aiguier@lami.univ-evry.fr⟩, University of Evry

Daniel Amyot ⟨damyot@site.uottawa.ca⟩, SITE, University of Ottawa

Karim Berkani ⟨karim.berkani@rd.francetelecom.fr⟩, France Telecom R&D

Lynne Blair ⟨lb@comp.lancs.ac.uk⟩, Lancaster University

Jan Bredereke ⟨brederek@tzi.de⟩, University of Bremen TZI

Wiet Bouma ⟨L.G.Bouma@kpn.com⟩, KPN Research

Franck Cassez ⟨Franck.Cassez@ircyn.ec-nantes.fr⟩, IRCyN Nantes

Bernard Cohen ⟨B.Cohen@city.ac.uk⟩, City University, London

Tianbao Ding ⟨tding@site.uottawa.ca⟩, SITE, University of Ottawa

Christophe Gaston ⟨gaston@lami.univ-evry.fr⟩, University of Evry

Stephen Gilmore ⟨stg@dcs.ed.ac.uk⟩, University of Edinburgh

Nikos Gorogiannis ⟨nkg@cs.bham.ac.uk⟩, University of Birmingham

Tom Gray ⟨tom_gray@mitel.com⟩, Mitel Corporation

Mario Kolberg ⟨mkolberg@acm.org⟩, University of Strathclyde

Pascale Le Gall ⟨legall@lami.univ-evry.fr⟩, University of Evry

Luigi Logrippo ⟨luigi@site.uottawa.ca⟩, SITE, University of Ottawa

Dominique Méry ⟨Dominique.Mery@loria.fr⟩, LORIA

Dave Marples ⟨dmarples@research.telcordia.com⟩, Telecordia Technologies

Gustaf Naeser ⟨gaffe@docs.uu.se⟩, Uppsala University

Jan Nyström ⟨jann@docs.uu.se⟩, Uppsala University

Tadashi Ohta ⟨ohta@t.soka.ac.jp⟩, Soka University

Farid Ouabdesselam ⟨Farid.Ouabdesselam@imag.fr⟩, Laboratoire LSR-IMAG

Jianxiong Pang ⟨j.pang@lancaster.ac.uk⟩, University of Lancaster

Ioannis Parissis ⟨Ioannis.Parissis@imag.fr⟩, Laboratoire LSR-IMAG

Malte Plath ⟨M.C.Plath@cs.bham.ac.uk⟩, University of Birmingham

Stephan Reiff ⟨sreiff@dcs.gla.ac.uk⟩, University of Glasgow

Jean-Luc Richier ⟨Jean-Luc.Richier@imag.fr⟩, Laboratoire LSR-IMAG

Mark Ryan ⟨mdr@cs.bham.ac.uk⟩, University of Birmingham

Pierre-Yves Schobbens ⟨pys@info.fundp.ac.be⟩, University of Namur

Jacques Sincennes ⟨jack@site.uottawa.ca⟩, SITE, University of Ottawa

Joanna Tomasik ⟨joto@dcs.ed.ac.uk⟩, University of Edinburgh

Ken Turner ⟨kjt@cs.stir.ac.uk⟩, Stirling University

Pamela Zave ⟨pamela@research.att.com⟩, AT&T Labs

Structuring Telecommunications Features

Kenneth J. Turner

Department of Computing Science and Mathematics, University of Stirling
Stirling FK9 4LA, Scotland (Email *kjt@cs.stir.ac.uk*)

Abstract. It is argued that languages constructs for defining services are important in providing well-structured descriptions. This view is illustrated with two
service definition languages, ANISE (Architectural Notions In Service Engineering)
and CRESS (CHISEL Representation Employing Structured Specifications). ANISE is
a mainly textual notation for defining services through the composition of simpler
features, right down to the most elementary behaviours. CRESS is a mainly graphical notation for defining services through the composition of features with a root
description. Both approaches are described briefly with short examples.

1 Introduction

The range of telecommunications services continues to grow rapidly. The
primary motivation seems to be that services differentiate providers, and
are significant sources of revenue in their own right. The terms service and
feature tend to be used loosely and interchangeably, although service should
be reserved for what is marketed to a user, and feature should be reserved
for a component of a service.

The IN (Intelligent Network [6–8]), AIN (Advanced Intelligent Network)
and TINA (Telecommunications Intelligent Network Architecture [3]) have
provided a framework for new telecommunications services. New kinds of
telecommunications services are being introduced in the multimedia, mobile
and wireless arenas. New ways of providing third-party services are emerging due to industry initiatives like JAIN (Java for Advanced/Java API for
Intelligent Networks [12]), Parlay [11] and TAPI (Telephony Application Programming Interface [10]). Service scripting by users (in the broadest sense)
is leading to new kinds of services.

There are two difficulties with this proliferation of services. Although each
initiative offers an approach to developing services, there is no uniform and
consistent architecture. The almost uncontrolled growth of services leads to
undesirable interference among services – the feature interaction problem [2].

It is the author's belief that telecommunications services urgently need
a theory and a uniform architecture. The title of this volume reflects the
same interest. There is considerable effort in the research community to detect feature interactions. This is, however, somewhat back-to-front. Consider
programming as an analogy. Programmers are expected to use good methods
in their program design and coding, for example structured/object-oriented
analysis and design. The emphasis is on developing well-structured programs.

Program debugging and testing remain necessary tasks, but these are hopefully diminished in importance by appropriate design techniques. By analogy, good structuring of services should be the main priority. Interaction detection should be treated as a lesser activity like debugging. Appropriate methods for structuring services can help to avoid interactions by design.

Structuring of services, and thus constructs for describing services, are thus crucial. The IN made progress towards this, but is rather dominated by bottom-up, engineering concerns rather than top-down, service concerns. This is borne out by the relative levels of sophistication in the service plane and the physical plane of the IN. The IN offers different approaches for describing services in the global functional plane and the distributed functional plane. The notion of SIBs (Service-Independent Building Blocks) is in principle a good solution. Unfortunately the SIBs defined for the IN are rather *ad hoc* in nature. They have not been justified as necessary or sufficient for the construction of the intended range of IN services.

SCEs (Service Creation Environments) are a similar but more pragmatic approach to structuring services. A number of companies have produced proprietary SCEs, e.g. Ericsson, Hewlett-Packard, Marconi Communications, Nortel and Siemens. The emphasis is again on building blocks, but these tend to be strongly oriented towards a particular manufacturer's equipment. This is hardly surprising since the aim is to generate code directly. As a result, the service descriptions have limited portability among manufacturers and are not as generic as they might be. SCEs are also focused on implementation rather than analysis. It would be desirable to analyse services as they are defined, well in advance of any implementation commitment.

The work reported here started from the basics of telecommunications services, with the intention of defining a uniform way of constructing services. The starting point in this approach is intentionally very low-level: the patterns of basic communication between a user and a service. Features and services are then built from these simple communications. The viewpoint is that of a service user, and deliberately omits as much as possible of the network internals. This is partly because the user view is the important one from a service perspective. It is also partly a matter of necessity, because an enormous amount of engineering detail is necessary to realise a network. Such minutiae would make service descriptions unworkably large, and would hide the essential aspects.

The outcome of the work is two complementary approaches: ANISE and CRESS. Although these are focused on IN-like services, they are not bound to the IN. For example, ANISE derives from another approach called SAGE (Service Attribute Generator [13]) used to define services in OSI (Open Systems Interconnection [5]). In particular, the basic call is *constructed* in ANISE and CRESS. This ensures that they can deal with other kinds of services, and not just those in telecommunications. There is insufficient space here to describe ANISE and CRESS in detail; see the referenced papers for more information.

2   ANISE

ANISE (Architectural Notions In Service Engineering [14–17,19]) constructs services through progressive combination of simpler behaviours. In the ANISE approach, there is no distinction between the simplest behaviours and the most complex services. All are termed 'features', though this does not match the normal usage of the word so well. However the terminology is logical in that the ability to dial a number, display a caller's number, or support three-way calls are simply different scales of behaviour.

The simplest behaviours in ANISE are called elementary features. These characterise the basic behaviour of the user-service interface. For example:

declare(Dial,
    feature(**12**,**local_confirmed**,**single**,Dial(Num),Dial(CallingMess)))

defines dialling a number. The **declare** construct names (as *Dial*) some piece of behaviour (here defined as an elementary feature). The words shown in bold are reserved in ANISE. Other identifiers are either user-defined (like *Dial*) or appear in the ANISE library definitions (like *Num* and *CallingMess*). In general, each ANISE declaration expands to a behaviour definition using macro-like mechanisms. Combinators take behaviour arguments and return their composite behaviour.

A **feature** declaration gives the direction, behaviour pattern, ordering property and primitives of a simple behaviour. In the above declaration of dialling, the direction is from the first user in a call towards the second (**12**). The dialling request is confirmed by the network back to the first user (**local_confirmed**), and occurs just once in a call (**single**). The dialling request is made by the *Dial* primitive with number parameter *Num*. The confirmation of this is also a *Dial* primitive with parameter *CallingMess*, meaning a response to a calling user. The effect of the declaration is to define the behaviour of dialling based on its abstract properties. Other elementary features in a basic call include seizing the line, ringing the called party's number, and clearing (hanging up) the call.

These behavioural 'atoms' are then composed into more complex behavioural 'molecules'. A small set of well-defined combinators (about 20) is sufficient for defining a wide range of telecommunications services. For example **duplicates**(*Speak*) takes the elementary feature for one-way speech and interleaves its behaviour in both directions, thus describing normal two-way speech. More complex combinators take pairs of behaviours. An example is **enables**(*Clear,Silence*), meaning that completion of clearing while ringing a number will stop ringing.

Some combinators describe conditional behaviour. As an example, the **interrupts_after_try** combinator allows behaviour to be interrupted only after it has been tried; for example, a user cannot clear a call until after a call attempt has begun. The **enables_on_result** combinator allows behaviour to continue only if it achieves some particular result; for example, dialling is possible only if seizing the line yields dialling tone. With these elementary features and combinators, a description of the basic call can be given compactly. ANISE is a textual language, but figure 1 shows graphically how it is used to structure the description of POTS (Plain Old Telephone Service). The inner circles represent elementary features such as the caller hanging up (*Clear1*) or the other party answering a call (*Answer*). The outer ovals represent combinations of simpler behaviours.

What would normally be regarded as features are changes to the root (POTS) description. These alterations are encapsulated in service-specific combinators. Calling Number Delivery, for example, changes ringing to include the caller's number; in ANISE terms, the **rings_display** combinator modifies ringing behaviour in this way. Call Forwarding on Busy Line uses the **diverts_busy** combinator that modifies the busy check for a line. Some features, such as Call Waiting and Three-Way Calling, are complex and need auxiliary definitions that use the generic constructs of ANISE.

ANISE is supported by a second language ANGEN (ANISE Generator) that automates the procedure for combining features with a root description. Although this has been exploited for telecommunications services, the approach is generic and could be used in other contexts. For example, a web browser could be regarded as providing a basic service that is modified by plug-ins. ANGEN defines the 'deltas' of a feature relative to the root description. This is achieved syntactically by stating how portions of the root description are extended, changed or deleted by the feature.

Although unrelated to feature construction, ANISE is supported by a third language ANTEST (ANISE Test) for definition of 'use-case scenarios' that characterise the expected behaviour of a feature. Proving correctness of features is in general very difficult or computationally intractable. ANTEST takes the more pragmatic approach of validating feature descriptions against a set of critical scenarios. ANTEST can be used to validate features in isolation. More importantly it is used to validate features in combination, i.e. to detect feature interactions. The ANISE interpretation of interaction is that a feature in isolation behaves differently when combined with others. In fact, ANISE allows features to be checked in arbitrary combinations (including all features together). Although ANTEST has been demonstrated only on telecommunications services, its approach is generic and could be used with other kinds of services.

ANISE is thus a collection of languages, techniques and tools. Descriptions are internally translated to LOTOS (Language Of Temporal Ordering Specification [4]). This ensures that ANISE descriptions have a well-defined meaning, and allows automated analysis to be performed. LOTOS is intentionally hidden in the approach because engineers in industry are generally unfamiliar with formal methods and are often reluctant to adopt them. The ANISE user therefore sees only the telecommunications-oriented feature descriptions and their analyses.

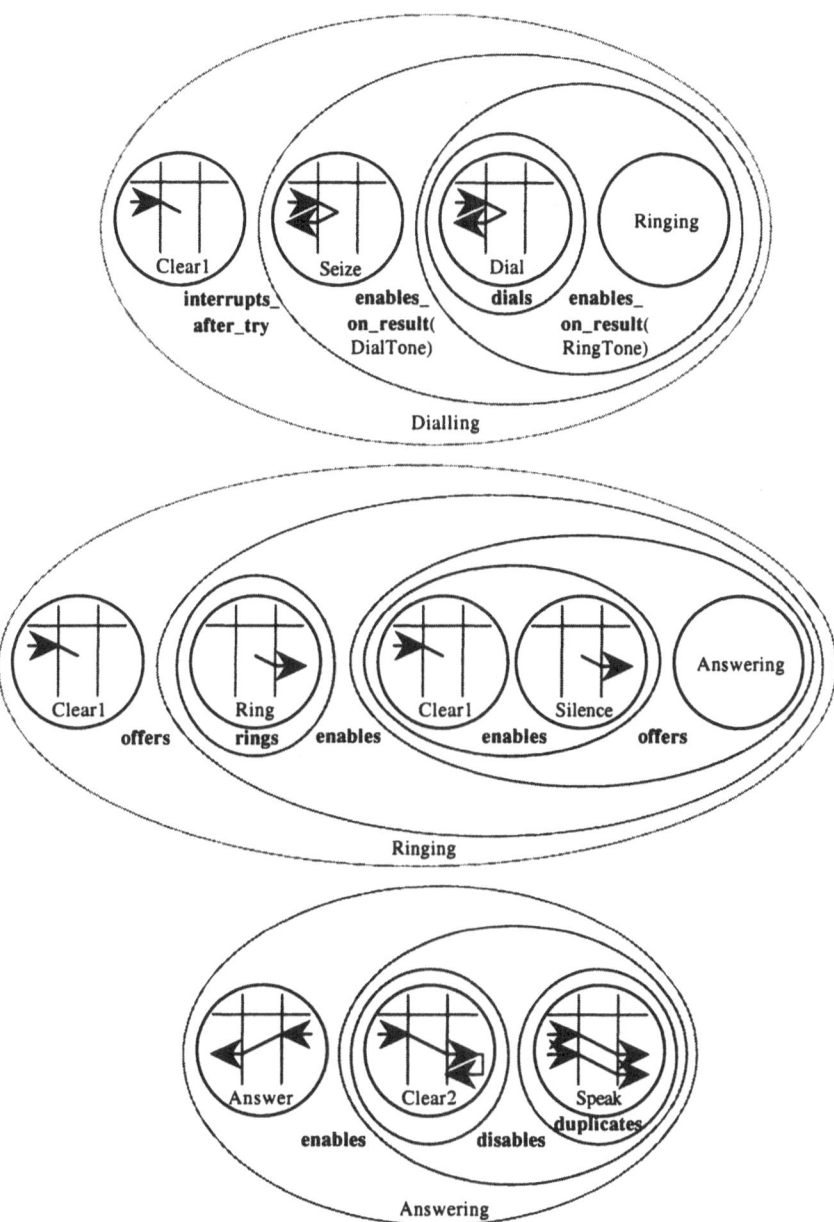

Fig. 1. Graphical Overview of ANISE Structure for POTS

3 CRESS

CHISEL [1] was developed by BellCore (now Telecordia Technologies) as a graphical notation for describing telecommunications services and features. It offers descriptions that can be readily assimilated by non-specialists, and yet can be translated to formal descriptions in various languages. CHISEL is attractive because it reflects industrial practice in defining features, and because it requires simple structuring mechanisms. However the rules for CHISEL are relatively informal. The author has therefore tightened up and extended the definition of CHISEL. The result is CRESS (CHISEL Representation Employing Structured Specifications [18]) – a feature notation backwards-compatible with CHISEL, but with a formalised interpretation using LOTOS and SDL (Specification and Description Language [9]).

CRESS starts with even simpler behaviours than ANISE: the isolated events that describe communication between a user and the network. These events are then connected using simple constructs such as sequence, guarded choice, iteration and parallel composition. Complex diagrams may be split into a number of interconnected parts.

Figure 2 shows POTS as a typical root service. In a different domain, a root diagram might describe some completely different kind of service. Some events are initiated by the subscriber (e.g. *Off-hook*) while others are initiated by the network (e.g. *DialTone*). Events carry parameters such as the calling and called numbers (A and B). The rounded rectangle at the top left of the figure defines various rules. Features have call variables that may be local to the feature (e.g. the calling and called numbers A and B) or may be global (e.g. *Busy* or *Ringing*). The rules for manipulation of these variables are formalised as assignments invoked under particular conditions. For example when some subscriber P goes off-hook (*Off-hook P*), this causes the line to be marked as busy (*Busy P* $\Leftarrow$ *True*).

Feature diagrams define how the root service is modified. This is done syntactically by describing how a feature is merged with the root. The following uses the example of CND (Calling Number Delivery), as shown in figure 3. These alternative diagrams build on the POTS description in figure 2. A feature is defined by adding, modifying or removing behaviour in the root diagram. The left-hand diagram in figure 3 modifies behaviour leading from source node *POTS 3*. The behaviour in node *POTS 4* is replaced by similar behaviour that checks whether or not calling number delivery is needed (*CallingNumber B*). If so, the calling number is displayed. In either case, the feature continues with the rest of the root diagram (target node *POTS 5* or *POTS 13*). The effect is to splice the CND feature into the POTS diagram. The right-hand description of CND in figure 3 is more convenient. Its first node begins with an asterisk, indicating a generic (macro-like) feature. It causes any diagram node containing *StartRinging* to be followed by a check for calling number delivery.

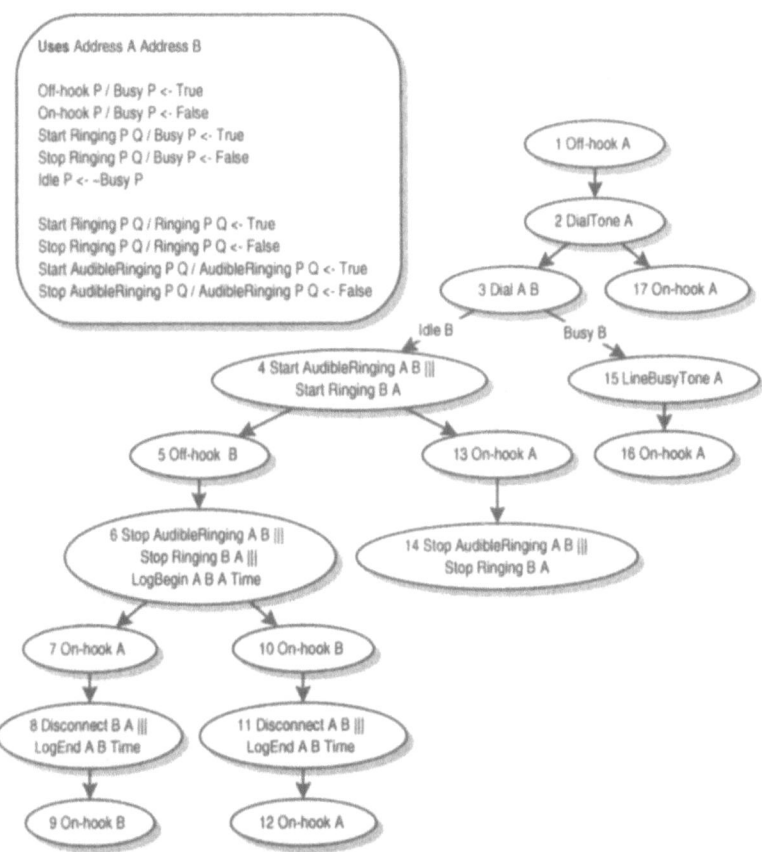

Fig. 2. Plain Old Telephone Service in CRESS

Features may be described in either of the ways shown in figure 3. The left-hand style is appropriate when a feature is to be spliced between particular nodes of a root diagram. The right-hand style is appropriate for defining a feature without knowledge of how the root diagram is organised (e.g. its specific nodes). Furthermore, this style allows a feature to be automatically instantiated a number of times. For example, Three-Way Calling has several instances of ringing where CND should be applied.

CRESS formalises the syntactic and static semantic rules for diagrams and their composition. It automates the combination of features and the root service. The composite diagram is then translated to LOTOS or SDL. This gives a formal representation of the behaviour that can be analysed by conventional means. For example, simulation, state exploration or model-checking can be used to find problems. Equivalence can be studied, and compatibility of a feature can be confirmed when composed with other features. As with ANISE, feature interaction is interpreted in CRESS as a change in feature behaviour when combined with other features.

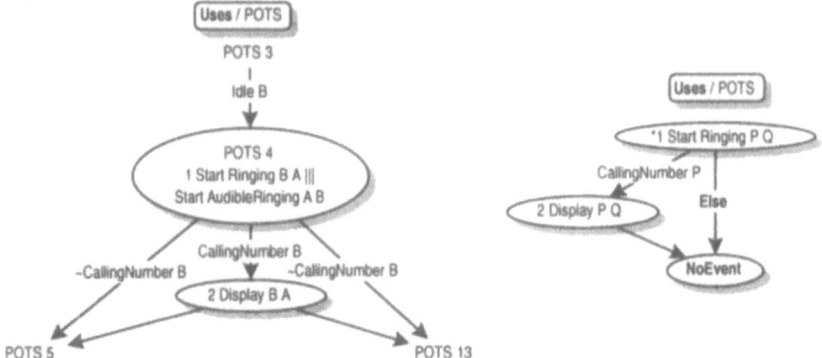

Fig. 3. Calling Number Delivery in CRESS

4   Conclusion

It has been argued that, as for programming, good structuring techniques are crucial when defining services. The ANISE and CRESS approaches have been briefly illustrated with small examples. Although ANISE and CRESS have been developed in the context of telecommunications services, future work will investigate their use on a wider range of services – especially from outside the telecommunications world.

References

1. Alfred V. Aho, Sean Gallagher, Nancy D. Griffeth, Cynthia R. Schell, and Deborah F. Swayne. SCF3/Sculptor with Chisel: Requirements engineering for communications services. In Kristofer Kimbler and Wiet Bouma, editors, *Proc. 5th. Feature Interactions in Telecommunications and Software Systems*, pages 45–63. IOS Press, Amsterdam, Netherlands, September 1998.
2. E. Jane Cameron, Nancy D. Griffeth, Yow-Jian Lin, Margaret E. Nilson, William K. Schnure, and Hugo Velthuijsen. A feature-interaction benchmark for IN and beyond. *IEEE Communications Magazine*, pages 64–69, March 1993.
3. Fabrice Dupuy, Gunnar Nilsson, and Yuji Inoue. The TINA consortium: Towards networking telecommunications information services. *IEEE Communications Magazine*, pages 78–83, November 1995.
4. ISO/IEC. *Information Processing Systems – Open Systems Interconnection – LOTOS – A Formal Description Technique based on the Temporal Ordering of Observational Behaviour*. ISO/IEC 8807. International Organization for Standardization, Geneva, Switzerland, 1989.
5. ISO/IEC. *Information Processing Systems – Open Systems Interconnection – Basic Reference Model*. ISO/IEC 7498. International Organization for Standardization, Geneva, Switzerland, 1994.
6. ITU. *Intelligent Network – Q.120x Series Intelligent Network Recommendation Structure*. ITU-T Q.1200. International Telecommunications Union, Geneva, Switzerland, 1993.

7. ITU. *Intelligent Network – Q.121x Series Intelligent Network Recommendation Structure*. ITU-T Q.1210. International Telecommunications Union, Geneva, Switzerland, 1993.

8. ITU. *Intelligent Network – Q.122x Series Intelligent Network Recommendation Structure*. ITU-T Q.1220. International Telecommunications Union, Geneva, Switzerland, 1996.

9. ITU. *Specification and Description Language*. ITU-T Z.100. International Telecommunications Union, Geneva, Switzerland, 1996.

10. Microsoft. TAPI (Telephony Application Programming Interface). http://www.microsoft.com/communications/tapilearn30.htm, April 2000.

11. Parlay Consortium. Parlay. http://parlay.msftlabs.com/, April 2000.

12. Sun Microsystems. JAIN (Java for Advanced/Java API for Intelligent Networks). http://java.sun.com/products/jain/, April 2000.

13. Kenneth J. Turner. An engineering approach to formal methods. In André A. S. Danthine, Guy Leduc, and Pierre Wolper, editors, *Proc. Protocol Specification, Testing and Verification XIII*, pages 357–380, Amsterdam, Netherlands, June 1993. North-Holland. Invited paper.

14. Kenneth J. Turner. An architectural foundation for relating features. In Petre Dini, Raouf Boutaba, and Luigi M. S. Logrippo, editors, *Proc. 4th. International Workshop on Feature Interactions in Telecommunication Networks and Software Systems*, pages 226–241, Amsterdam, Netherlands, 1997. IOS Press.

15. Kenneth J. Turner. Relating services and features in the intelligent network. In Marijan Kunštić, editor, *Proc. 4th. International Conference on Telecommunications*, pages 235–243, Croatia, June 1997. University of Zagreb.

16. Kenneth J. Turner. An architectural description of intelligent network features and their interactions. *Computer Networks*, 30(15):1389–1419, September 1998.

17. Kenneth J. Turner. Validating architectural feature descriptions using LOTOS. In Kristofer Kimbler and Wiet Bouma, editors, *Proc. 5th. Feature Interactions in Telecommunications and Software Systems*, pages 247–261, Amsterdam, Netherlands, September 1998. IOS Press.

18. Kenneth J. Turner. Formalising the CHISEL feature notation. In Muffy H. Calder and Evan H. Magill, editors, *Proc. 6th. Feature Interactions in Telecommunications and Software Systems*, pages 241–256, Amsterdam, Netherlands, May 2000. IOS Press.

19. Kenneth J. Turner. Realising architectural feature descriptions using LOTOS. *Parallel Computers, Networks and Distributed Systems (Calculateurs Parallèles, Réseaux et Systèmes Répartis)*, September 2000.

Feature-Oriented Description, Formal Methods, and DFC

Pamela Zave

AT&T Research, Shannon Laboratory, Florham Park, New Jersey 07932, USA

Abstract. This paper explores the close relationship between feature-oriented and architectural styles of formal system description, and explains the formal methods needed to support descriptions in these styles. In the telecommunications domain, DFC is a successful example of a feature-oriented, architectural description technique. Although formal methods for DFC are in their infancy, there are already simple examples of all the necessary methods.

1 Feature-Oriented Description

1.1 Features

A *feature* of a software system is an optional or incremental unit of functionality. A *feature-oriented description* is a description of a software system organized by features. It consists of a base description and feature modules, each of which describes a separate feature. The set of possible system behaviors is determined by applying a feature-composition operator to the base description and these modules.

The big attraction of feature-oriented description is *behavioral modularity*, which makes it possible to change the system's behavior easily. With *perfect* behavioral modularity, it would be possible to make *any* desired change to the behavior of a system by composing a new feature module with the existing system description. It would never be necessary to change existing modules, yet the compositional structure of the description would make it comprehensible and analyzable overall.

Feature-oriented description has become widely popular, particularly in application domains such as telecommunications with large, long-lasting systems and many diverse stakeholders. Continual evolution is the biggest challenge for telecommunications software, probably outweighing all the others combined.

Unfortunately, the vast majority of feature-oriented descriptions are informal—the feature modules are written in natural language, and the feature-composition operator is concatenation of the text. Any desired change to system behavior is easy to make: just describe the change in natural language, whatever it is. On the other hand, the description is neither comprehensible nor analyzable overall. It rarely defines the system's behavior in a complete, consistent, and unambiguous manner. This situation affects all segments of the telecommunication industry, and is the primary motivation for the industry's interest in formal methods [7–10,20].

The challenge for researchers in formal methods is to find a way to combine feature-oriented description with formal description, realizing to a degree useful for evolving systems the advantages of both. This is proving to be very difficult. The history of research on description of telecommunication systems [7–10,20] records many types of feature composition, each illustrated with a handful of features, but few that seem applicable to telecommunication systems of realistic size, with hundreds or even thousands of features.

1.2   Feature Interaction

A *feature interaction* is some way in which a feature or features modify or influence another feature in describing the system's behavior set. Formally this influence can take many forms, depending on the nature of the feature-description language and composition operator. A group of logical assertions, composed by conjunction, can affect each other's meanings rather differently than a group of finite-state machines, composed by event synchronization.

In general, features might interact by causing their composition to be incomplete, inconsistent, nondeterministic, or unimplementable in some specific sense.[1] Or the presence of a feature might simply change the meaning of another feature with which it interacts.

Feature-oriented description emphasizes individual features and makes them explicit. It also de-emphasizes feature interactions, and makes them implicit in the effects of the feature-composition operator. First and foremost, formal methods for feature-oriented system descriptions must help engineers manage feature interactions.

Two points about feature interactions are frequently misunderstood. Since these misunderstandings make it impossible to talk about feature interaction clearly, let alone formally, it is important to emphasize these points:

- While many feature interactions are undesirable, many others are desirable or necessary. *Not all feature interactions are bad!*
- Feature interactions are an inevitable by-product of behavioral modularity.

"Busy treatments" in telephony are features for handling busy situations. They exemplify these points. Suppose that we have a feature-description language in which a busy treatment is specified by providing an action, an enabling condition, and a priority. Further suppose that a special feature-composition operator ensures that, in any busy situation, the action applied will be that of the highest-priority enabled busy treatment.

---

[1] For example, in TLA [1] the result of feature composition could fail to be *machine-closed*. The specification would be unimplementable because it requires the system to control the environment's choices.

In a busy situation where two busy treatments B_1 and B_2 are both enabled, with B_2 having higher priority, these features will interact: the action of B_1 will not be applied, even though its stand-alone description says that it should be applied. This feature interaction is intentional and desirable. It is a by-product of the behavioral modularity that allows us to add busy treatments to the system without changing existing busy treatments. Without the special composition operator, when B_2 is added to the system, the enabling condition E_1 of B_1 must be changed to $E_1 \wedge \neg E_2$.

1.3   Feature Engineering

Among the major artifacts of software development [13] are:

- *Domain knowledge,* which provides presumed facts about the environment.
- *Requirements,* which indicate what the stakeholders need from the system, described in terms of its desired effect on the environment.

The major proof obligation of formal methods for software development is to show that a description of the system, in conjunction with the domain knowledge, implies the satisfaction of the requirements.

Of course, this approach to establishing the adequacy of a system description is only as good as the requirements provided. Unfortunately, it is difficult to write complete requirements for a feature-oriented system, for two reasons:

- The behavior of a feature-oriented system changes, sometimes radically, with the addition of every feature. An assertion that is true of today's system but not true of tomorrow's has little value. Assertions that are likely to be true of all future systems are few and far between.
- To write complete requirements, it would be necessary to state formally and explicitly exactly how all features interact. This is something people tend to do poorly [23], and it is exactly the chore that feature-oriented description was invented to avoid.

Some requirements seem useful and feasible. For example, it is often possible to write good requirements for the behavior of an individual feature or small group of features. For another example, the system's behavior might have been designed from the beginning to follow unifying principles of etiquette and feature integration.

Inevitably, however, the set of all such requirements will constrain the system's behavior only loosely. For all the other aspects of the system's behavior, where feature interaction is the critical issue, we must use a different type of formal method called here *feature engineering*. Feature engineering is the process of engineering features and feature interactions for maximum user satisfaction and system integrity. Feature engineering develops the missing system requirements from the bottom up.

Figure 1 is a simple picture of how feature engineering might be done. It assumes that a base specification and old features already exist, and that the goal is to add new features to the system.

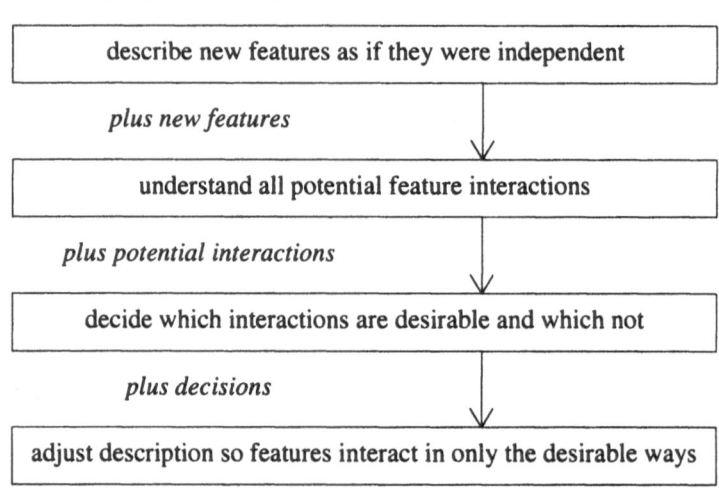

Fig. 1. A proposed process for feature engineering.

In the first step of the process, new features are described, to as great an extent as possible, as if they were independent of all others. People will want to do this anyway, as there is a strong human tendency to ignore feature interactions. They will be able to do it only if the description technique has sufficient behavioral modularity.

In the second step of the process, engineers must come to understand all potential feature interactions. This is essentially impossible for people to do without the help of automated analysis, based on a classification of the possible interactions and on algorithms for detecting their presence in a feature set.

In the third step of the process, engineers classify potential interactions as bad or good. This step is primarily manual. As engineers gain experience and insight, they might discover general principles that govern some of their choices. Such principles can feed into automated assistance for this step, or even be elevated to the status of requirements.

In the fourth step of the process, feature descriptions are adjusted so that all of the good feature interactions are present and none of the bad ones are. If the description technique has sufficient behavioral modularity, only the descriptions of new features need be changed, and all the old features can remain untouched.

1.4 Formal Methods for Feature-Oriented Descriptions

A recent reference model of the proof obligations for formal methods of software development ([13], based on [16,18,29]) proposes three proof obligations. The first proof obligation, loosely paraphrased, says that the domain knowledge must be consistent or satisfiable. This is in no way dependent on how the system is described, so there is nothing special to say about it in the context of feature-oriented system descriptions.

The second proof obligation, *very* loosely paraphrased, says that for every possible behavior of the environment, there must be a behavior of the system that is consistent with the system description.[2]

The third proof obligation says that a description of the system, in conjunction with the domain knowledge, must imply the satisfaction of the requirements. As explained above, in a feature-oriented context this obligation must be met using a combination of ordinary verification and feature engineering.

2 Architectural Descriptions

2.1 Definitions of Architectural Description and Specification

Software engineers can choose whether or not to use feature-oriented descriptions of system behavior. In another dimension, they can choose whether their system descriptions are *architectural descriptions* or *specifications*. The distinction between these two is based on the phenomena whose relationships they are describing.

As mentioned in Section 1.3, domain knowledge and requirements are descriptions of environment phenomena. Obviously a system description must be a description of system phenomena. The interface between system and environment consists of *shared phenomena* that are visible to both.

A *specification* is a description of the system's behavior, written strictly in terms of shared phenomena. Two other familiar artifacts of software development are [13]:

- The *programming platform* provides the basis for programming a system to satisfy the requirements.
- The *program* implements the requirements on the programming platform.

These descriptions are not confined to system phenomena that are shared with the environment. Some of the relationships among the five artifacts or descriptions are illustrated by Figure 2.

A specification can be a complete description of system behavior, as also can a combination of program and platform. An obvious disadvantage of

---

[2] The aspects of the proof obligation omitted by this paraphrase concern the description of the environment rather than the description of the system, which is why they are irrelevant here.

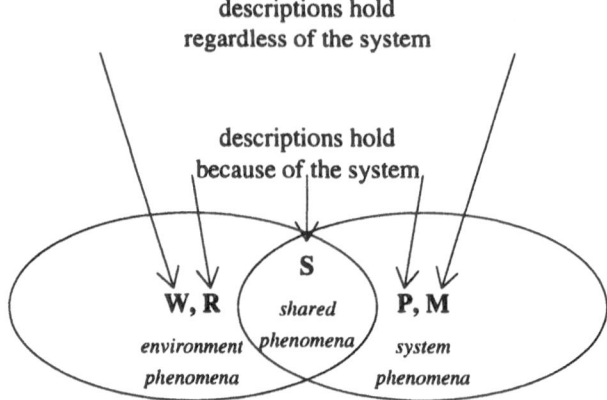

descriptions hold
regardless of the system

descriptions hold
because of the system

S

shared
phenomena

W, R

environment
phenomena

P, M

system
phenomena

Fig. 2. Some relationships among five descriptions [13]. W ("world") is domain knowledge, R is requirements, S is a specification, P is a program or programs, and M ("machine") is a programming platform.

describing system behavior in terms of program and platform is that it is difficult to reason about a low-level, general-purpose programming platform. Addressing this disadvantage, it is possible to describe a more abstract, idealized, domain-dependent programming platform. Then system behavior can be described in terms of this platform and the high-level, domain-specific programs written for it. Since an abstract programming platform is often thought of as an *architecture,* this combination can be termed an *architectural description.*

2.2   Assessment of Architectural Description and Specification

There are significant practical differences between architectural description and specification. In terms of the process and organization of software development, the major advantage of a specification is that it tells requirements engineers and stakeholders a minimum about the system, and implementors a minimum about the environment. It separates their concerns; it can act as a contract or a firewall between them. On each side of the wall engineers retain maximum freedom to do their job without consulting the other side, and are distracted by a minimum of information that is not significant to them.

In the same terms, the major advantage of architectural description is that it is straightforward to implement. It does not have the abstraction and separation of concerns that characterize a specification, but it is clearly—and maybe even efficiently—executable.

In terms of the actual construction of the system description, an architectural description is much easier to write than a specification (at least for realistically complex systems, if not for academically simple ones). The reason is that a specification must confine itself to a fixed set of phenomena; nothing that is not observable at the system/environment interface can be used.

Writers of architectural descriptions, on the other hand, can invent any extra phenomena that they find convenient. For example, internal system events are often useful. These events can be used to construct temporal sequences, cause-and-effect chains, and other helpful structures. The larger the description, the more important it is to have an organizing framework within which to place and locate the real information.

This is no secret to designers of specification languages, who often provide information hiding so that specifiers can use extra phenomena such as events, yet declare them to be optional in an implementation. However, the information-hiding approach imposes a very heavy burden on formal methods: to prove that a true implementation is behaviorally equivalent to the pseudo-implementation in the specification.

2.3   Feature-Oriented Description and Architectural Description: A Good Fit

From Section 1 we can conclude that feature-oriented descriptions are advantageous for evolving systems. The feature-oriented description technique must have behavioral modularity, and it must support three types of formal reasoning:

- reasoning that system behavior is well-defined in all possible situations;
- reasoning that assertions (requirements) on the system's environment are satisfied;
- structural classification and detection of potential feature interactions.

A system description can be both feature-oriented and architectural (Figure 3). The two styles fit together well, and there are several reasons for believing that using architectural description is the best way to achieve feature-oriented description.

The first reason is success. The proceedings of the Feature Interaction Workshops [7–10,20] show clearly that architectural descriptions have been the most successful in delivering behavioral modularity, and there is a definite trend toward architectural approaches [6,2,11,12,19,22,31]. One of the "helpful structures" that can be created with a free choice of internal system phenomena is behavioral modularity!

A second reason is that the architectural structure that defines and constrains feature composition is also extremely helpful in providing at least two of the three necessary types of formal reasoning. It can help show that system behavior is well-defined in all possible situations. For example, it might be used to prove that a component obeying all the rules of the architecture

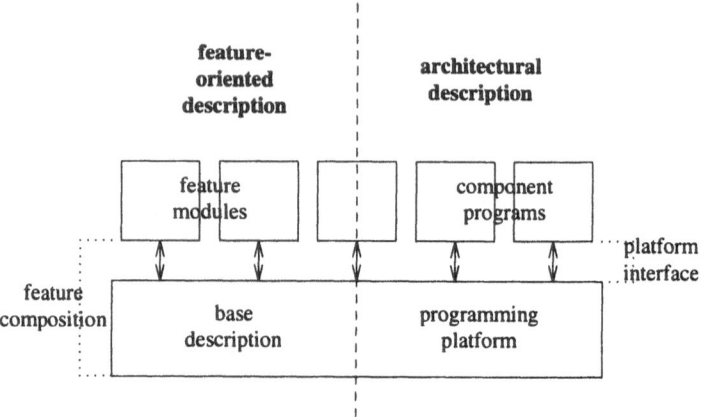

Fig. 3. The same system description can be viewed as both feature-oriented and architectural.

can never cause the system to deadlock. It is also essential in classifying and detecting potential feature interactions.

A third reason resides in the process of developing and managing an evolving system. One of the best ways to achieve timely evolution is to maintain an open system—one to which many parties contribute. Contributors to an open system are likely to be doing both requirements engineering and implementation, so they don't need a separation of those concerns. They do need a well-defined platform on which to program, one that elucidates the semantics of each component, helps different components cooperate within the overall system structure, and protects the integrity of the system from errant components.

Finally, if a formally described system is evolving, there is a good chance that it also interoperates with at least one legacy system that has not been developed with the help of formal methods. In this case, it is necessary to reverse-engineer a partial formal description of the legacy system, which is part of the environment of the newer system.

The newer system's interface with the legacy system will be different from its interface with users, and will expose system phenomena that are not shared with users. This is significant simply because of the extra convenience of architectural specification. Since implementation secrets are already being exposed by this extra interface, there is less motivation to go to the trouble of hiding them behind the wall of a specification.

All of the reasons for using architectural description apply to telecommunication systems. For many years, providers of telecommunication services have been demanding of vendors of telecommunication equipment that their systems be open. Service providers are frustrated with the slow pace of feature development by vendors, and want to speed it up by developing their own features or getting third parties to do it. Furthermore, any telecommunication system has an enormous legacy to deal with.

3   Distributed Feature Composition

3.1   The DFC Architecture

Distributed Feature Composition (DFC) is a new feature-oriented architecture for describing telecommunication services [17,28,27]. Its primary design goals were generality, analyzability, and behavioral modularity. It is now being implemented on an I.P. substrate within AT&T Research [4].

Regarding generality, hundreds of features and services have been described informally within the DFC framework, and we know of no services (including mobile and multimedia services) that cannot be fit into the architecture. The next section discusses analyzability. The following overview of DFC leaves out many details and even whole aspects; it is intended just to convey an impression of how DFC achieves behavioral modularity.

In DFC a customer call generates and is responded to by a *usage*, which is a dynamic assembly of *boxes* and *internal calls*. A *box* is a module, and implements either a line/device interface or a feature. An *internal call* is a featureless connection between two ports on two different boxes. Figure 4 illustrates a simple usage at two points in time.

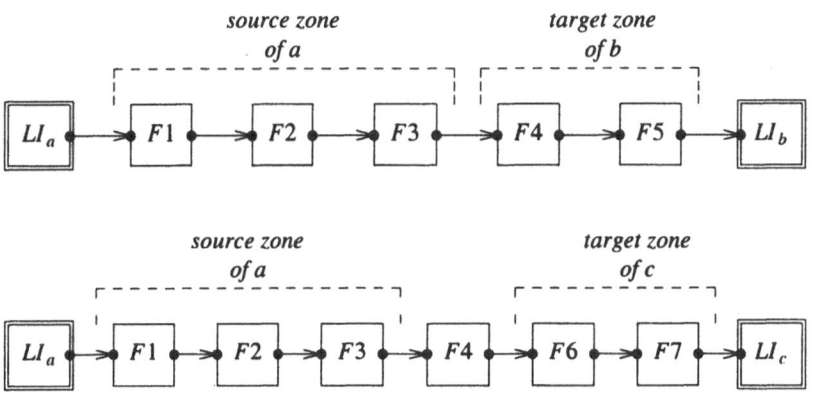

Fig. 4. Two snapshots of a linear usage.

In Figure 4 a DFC internal call is shown as an arrow from the port that placed the call to the port that received the call. Each internal call begins with a *setup phase* in which the initiating port sends a setup signal to the DFC router, and the DFC router chooses a box and forwards the signal to it. The receiving box chooses an idle port for the call (if there is one) and completes the setup phase with a signal back to the initiating port. From that time until the *teardown phase*, the call exists and has a two-way signaling channel. The call can also have any number of media channels, each carrying any medium. Because there is no default configuration of media channels in a call (for example, it is not necessary for a call to have a voice channel), each media channel in a call must be opened and closed explicitly by signals on the signaling channel.

Having full control of all the calls it places or receives, a feature box has the autonomy to fulfill its purpose without external assistance. When a feature box does not need to function, it can behave *transparently*. For a box with two ports, both of which are engaged in calls, transparent behavior is sending any signal received from one port out the other port, and connecting the media channels in both directions. The two calls will behave as one, and the presence of the transparent box will not be observable by any other box in the usage.

When its function requires, a feature box can also behave assertively, re-routing internal calls, processing media streams, and absorbing/generating signals. To give a simple example, a Call Forwarding on No Answer box (box F_4 in Figure 4) first makes an outgoing internal call to address b as shown in the upper part of Figure 4. If CFNA receives a signal through its outgoing call that b is alerting, and then no other signal for 30 seconds, CFNA tears down its outgoing internal call and places a new outgoing internal call whose target is the forwarding address c. The resulting usage is shown in the lower part of Figure 4.

Most components of the DFC architecture are shown in Figure 5. The line-interface boxes (LI) are connected to telecommunication devices by external lines. The trunk-interface boxes (TI) are connected to other networks by external trunks. The feature boxes (F) can have any number of ports, depending on their various functions. Internal calls are provided by the port-to-port *virtual network*.

The router of the virtual network is unusual. It not only routes internal calls to the destinations associated with their target addresses, as any network router does, but it also "applies features" by routing internal calls to and from feature boxes. For this reason it needs data on feature subscriptions, feature precedences, and the dialing plan as well as normal configuration data. (All global data is shown in double rectangles in Figure 5.)

Figure 5 also shows global data called *operational data*, which is used by feature boxes. For example, the CFNA box retrieves its subscriber's forwarding address from operational data. Access to operational data is strictly partitioned by features, so its use cannot compromise feature modularity. Operational data is used mainly to provide provisioned customer information to features.

Each box fits into one of two large categories. A *bound box* is a unique, persistent, addressable individual. Bound boxes are doubled in Figures 4

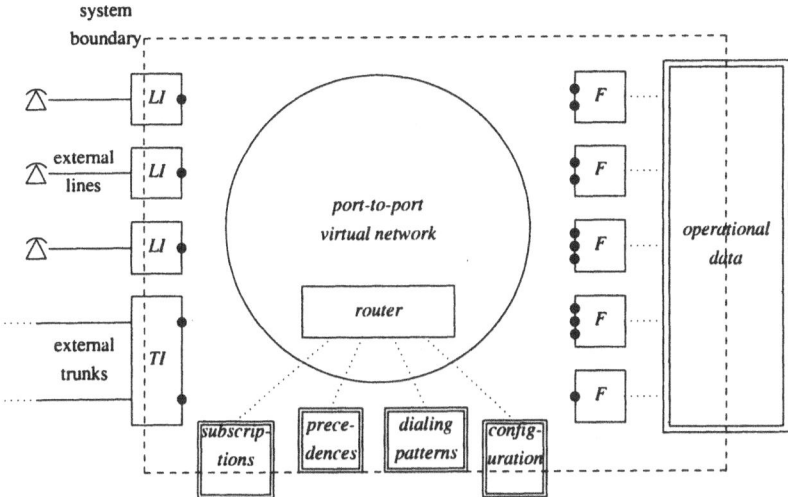

Fig. 5. Components of the DFC architecture.

and 6. In Figure 4 the only bound boxes are the two line interfaces. The other boxes in Figure 4 are *free boxes,* meaning that each box is an anonymous, interchangeable copy of its type with no persistence outside its tenure in a usage. The value of bound feature boxes is that they make it possible to have joins in usage graphs.

Figure 6 shows a usage in which there is a bound feature box representing Call Waiting. The CW_c box is associated with LI_c (which, like all the other line interfaces in this usage, is associated with an old-fashioned telephone). The usage was formed because first the customer owning address c made a successful call to address a. This passed transparently through CW_c just because c subscribes to CW. Later, customer b attempted to call c. An internal call generated by his attempt was routed to CW_c on its way to c, where it was accepted at the third port.

Acceptance of a call at its third port causes CW_c to spring into action. It first signals back to LI_b that the customer call has reached an alerting state; LI_b will play a ringback tone so that its customer can hear it. CW_c then alerts LI_c by inserting a tone on the voice channel to it. CW_c then monitors the voice channel from LI_c for flash signals. Each time CW_c recognizes a flash signal, it switches its internal voice path between the two possible positions: connecting the ports that connect c with a, or connecting the ports that connect c with b.

Figure 6 illustrates the important distinction between a usage and a customer call. There are two customer calls, one placed by customer c and one placed by customer b. But there is only one usage, representing the joining of

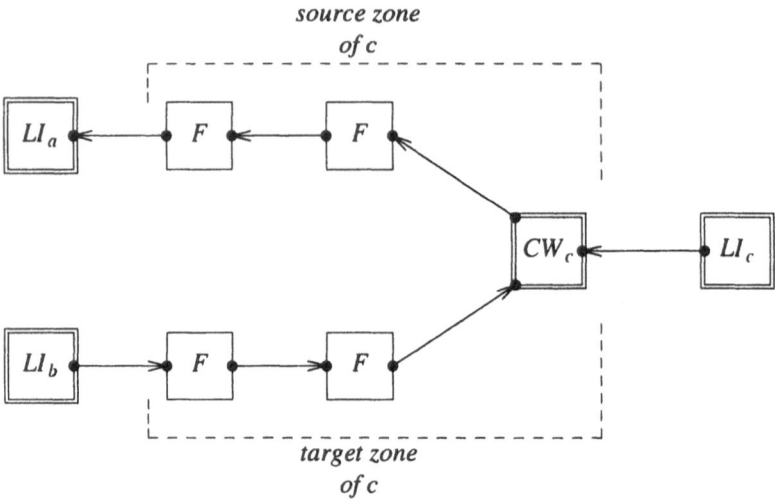

Fig. 6. A nonlinear usage.

those separate communication attempts by the CW feature. Many difficulties in telecommunications are due to ignoring this crucial distinction.

The key to assembly of the necessary usage configurations is the DFC routing algorithm. It operates on the setup signal that initiates each internal call. The setup signal has five fields of interest to the router; in the form **name**: *type* they are: **source**: *address*, **dialed**: *string*, **target**: *address*, **command**: {new, continue, update}, and **route**: **seq** *routing_pair*. Each routing pair has a first component of type *box_type* and a second component of type *zone* = { source, dialed, target }.

We shall explain the function of the router by example, first describing how the usage in Figure 4 evolved. The setup signal emitted by LI_a had a **source** field containing a, a **dialed** field containing the dialed string, and a **command** field containing **new**. The other two fields were empty. Upon receiving the signal, the router first extracted the target line address b from the dialed string, and put it into the **target** field.

Next the router, instructed by **new**, computed a new route and put it into the **route** field. Customer a subscribes to three features F1, F2, and F3 in the source zone, so the first three pairs of the route were *(F1*,source*)*, *(F2*,source*)*, and *(F3*,source*)*. If the dialed string had matched the triggering pattern of one or more features in the dialed zone, routing pairs for those features would have been next, but this was not the case so there were no routing pairs for dialed-zone features. The target address b subscribes to two features F4 and F5 in the target zone, so the last two pairs of the route were *(F4*,target*)* and *(F5*,target*)*.

Now the router had finished manipulating the setup signal, and needed to find a box to route the internal call to. It stripped the first pair off the route, and since F1 is the type of a free box, it routed the internal call to an arbitrary fresh box of that type.

The feature boxes in the upper part of Figure 4 had no initial need to control the routing. So when each box prepared a setup signal for an outgoing call, it simply copied the entire setup signal from its incoming call, making sure that the command field had continue rather than new. The continue command told the router not to recompute anything in the route. The chain unfolded, one pair of the route being deleted as each free box was added to the usage. Finally, in the last internal call, the route was empty so the router routed to the bound box LI_b.

When the CFNA feature box ($F4$) made its second outgoing call, the value of target in the setup signal was the forwarding address c. To ensure correct feature application, CFNA set the command value to update(target). This caused the router to remove from the route the remnants of the target zone of b, and to replace them with a newly computed target-zone route for c. Because of this substitution the usage was routed through the target features of c before reaching LI_c, as shown in the lower part of Figure 4.

In constructing a route, the router uses a *precedence* relation governing the order in which features can occur in a route (precedences are the only place in a DFC system where features are related explicitly to one another). This order, of course, has important effects on how features interact. For example, a busy signal usually originates in the target interface box. So the proper coordination of busy treatments mentioned in Section 1.2 would be achieved by placing the higher-priority feature boxes later in the route, i.e., closer to the source of their triggering signal. A busy treatment absorbs and responds to a busy signal if its feature is enabled, and forwards the signal toward the earlier feature boxes if it is not enabled.

A user of CW must subscribe to CW in both the source and target zones, to ensure that every relevant communication goes through the same box. In Figure 3 the customer call from c to a goes through CW_c because c subscribes to CW (early) in the source zone. The customer call from b to c goes through CW_c because c subscribes to CW (late) in the target zone. Because CW is the type of a bound box, and CW_c is the unique box of this type associated with LI_c, both customer calls are routed to exactly the same box.

The major source of modularity in DFC is that features communicate with each other only through DFC internal calls. A feature box does not know what is on the other end of its calls, for example, whether a far port is associated with a feature or a user. So the feature box need not change when its environment changes, for example by the addition of another feature. At the same time, all the relevant signaling and media channels go through the feature box, so the feature box has the power to manipulate them in any way it finds necessary.

The DFC architecture is actually a domain-specific adaptation of the pipes-and-filters architecture [21]. Its modularity is exactly the same kind as that claimed for pipes and filters in general.

3.2   Formal Methods for DFC

Although the development of formal methods and tools for DFC is in its infancy, there have already been some interesting results.

The DFC architecture has been defined formally in a combination of Promela and Z [25]. The internal protocols of the architecture have been checked extensively [28] using Spin, a model checker for Promela [14].

As explained in Section 2.3, the obligation to show that system behavior is well-defined in all possible situations is best approached through architectural structure. For example, a tool has been built [5] to check that individual box programs written in a Statechart-like language [3] obey the DFC internal protocols and have well-defined behavior in all possible situations allowed by the DFC internal protocols. Inductive reasoning can be applied to show that if all box programs are vetted by the tool, the system cannot deadlock.

There is little experience so far with verifying true requirements. However, as one example, Daniel Jackson and I have proved an assertion using the Alloy Constraint Analyzer [15]. The assertion is a requirement for the behavior of multiple instances of a single complex feature, all of which must cooperate to achieve a particular result.

I have done one substantial study concerning the engineering of a particular class of feature [24,26]. The possible interactions among features of this class are categorized, and an efficient algorithm is proposed to detect them in a feature set. The entire process shown in Figure 1 is applied to an example set of these features.

There is a growing body of evidence that DFC offers a useful degree of behavioral modularity. In the study above, it was possible to achieve exactly the desired feature interactions without changing any feature modules, rather by simply adjusting the precedence relation used in DFC routing. In a study of interactions among mobile and multimedia features [30], it proved possible to describe neatly in DFC many features and feature interactions that had not previously been recognized.

References

1. Martín Abadi and Leslie Lamport. Composing specifications. *ACM Transactions on Programming Languages and Systems* XV(1):73-132, January 1993.
2. Pansy K. Au and Joanne M. Atlee. Evaluation of a state-based model of feature interactions. In [10], pages 153-167.
3. Gregory W. Bond. ECLIPSE Statecharts: A language for the design and implementation of telecommunication services. AT&T Laboratories Technical Report, June 2000.

4. Greg Bond, Eric Cheung, Andrew Forrest, Michael Jackson, Hal Purdy, Chris Ramming, and Pamela Zave. DFC as the basis for ECLIPSE, an IP communications software platform. In *Proceedings of the IP Telecom Services Workshop 2000*, to appear.

5. Greg Bond, Franjo Ivancic, Nils Klarlund, and Richard Trefler. Personal communication on Eclipse2Mocha.

6. Kenneth H. Braithwaite and Joanne M. Atlee. Towards automated detection of feature interactions. In [7], pages 36-57.

7. L. G. Bouma and H. Velthuijsen, editors. *Feature Interactions in Telecommunications Systems*. IOS Press, Amsterdam, 1994.

8. M. Calder and E. Magill, editors, *Feature Interactions in Telecommunications and Software Systems VI.*, IOS Press, Amsterdam, 2000.

9. K. E. Cheng and T. Ohta, editors, *Feature Interactions in Telecommunications Systems III.*, IOS Press, Amsterdam, 1995.

10. P. Dini, R. Boutaba, and L. Logrippo, editors. *Feature Interactions in Telecommunication Networks IV*. IOS Press, Amsterdam, 1997.

11. José M. Duran and John Visser. International standards for intelligent networks. *IEEE Communications* XXX(2):34-42, February 1992.

12. James J. Garrahan, Peter A. Russo, Kenichi Kitami, and Roberto Kung. Intelligent Network overview. *IEEE Communications* XXXI(3):30-36, March 1993.

13. Carl A. Gunter, Elsa L. Gunter, Michael Jackson, and Pamela Zave. A reference model for requirements and specifications. *IEEE Software* XVII(3):37-43, May/June 2000.

14. Gerard J. Holzmann. Spin web site at `http://netlib.bell-labs.com/netlib/spin/whatispin.html`.

15. Daniel Jackson. Alloy Constraint Analyzer web site at `http://sdg.lcs.mit.edu/alloy`.

16. Michael Jackson and Pamela Zave. Deriving specifications from requirements: An example. In *Proceedings of the Seventeenth International Conference on Software Engineering*, pages 15-24. ACM Press, ISBN 0-89791-708-1, 1995.

17. Michael Jackson and Pamela Zave. Distributed feature composition: A virtual architecture for telecommunications services. *IEEE Transactions on Software Engineering* XXIV(10):831-847, October 1998.

18. Michael Jackson and Pamela Zave. Domain descriptions. In *Proceedings of the IEEE International Symposium on Requirements Engineering*, pages 56-64. IEEE Computer Society Press, ISBN 0-8186-3120-1, 1992.

19. Jalel Kamoun and Luigi Logrippo. Goal-oriented feature interaction detection in the Intelligent Network model. In [20], pages 172-186.

20. K. Kimbler and L. G. Bouma, editors. *Feature Interactions in Telecommunications and Software Systems V*. IOS Press, Amsterdam, 1998.

21. Mary Shaw and David Garlan. *Software Architecture: Perspectives on an Emerging Discipline*. Prentice-Hall, 1996.

22. Greg Utas. A pattern language of feature interaction. In [20], pages 98-114.

23. Hugo Velthuijsen. Issues of non-monotonicity in feature-interaction detection. In [9], pages 31-42.

24. Pamela Zave. An experiment in feature engineering. In *Essays by the Members of the IFIP Working Group on Programming Methodology*, Springer-Verlag, to appear.

25. Pamela Zave. Formal description of telecommunication services in Promela and Z. In Manfred Broy and Ralf Steinbrüggen, editors, *Calculational System Design (Proceedings of the Nineteenth International NATO Summer School)*, pages 395-420. IOS Press, ISBN 90-5199-459-1, 1999.
26. Pamela Zave. Systematic design of call-coverage features. AT&T Laboratories Technical Report, November 1999.
27. Pamela Zave and Michael Jackson. DFC modifications I (Version 2): Routing extensions. AT&T Laboratories Technical Report, January 2000.
28. Pamela Zave and Michael Jackson. DFC modifications II: Protocol extensions. AT&T Laboratories Technical Report, November 1999.
29. Pamela Zave and Michael Jackson. Four dark corners of requirements engineering. *ACM Transactions on Software Engineering and Methodology* VI(1):1-30, January 1997.
30. Pamela Zave and Michael Jackson. New feature interactions in mobile and multimedia telecommunication services. In [8], pages 51-66.
31. Israel Zibman, Carl Woolf, Peter O'Reilly, Larry Strickland, David Willis, and John Visser. Minimizing feature interactions: An architecture and processing model approach. In [9], pages 65-83.

Use Case Maps as a Feature Description Notation

Daniel Amyot

Mitel Corporation, Kanata, Canada, and
School of Information Technology and Engineering, University of Ottawa
150 Louis-Pasteur, Ottawa, Ontario, K1N 6N5, Canada
Email: damyot@site.uottawa.ca
Web: http://www.UseCaseMaps.org

Abstract. We propose Use Case Maps (UCMs) as a notation for describing features. UCMs capture functional requirements in terms of causal scenarios bound to underlying abstract components. This particular view proved very useful in the description of a wide range of reactive and telecommunications systems. This paper presents some of the most interesting constructs and benefits of the notation in relation to a question on a User Requirements Notation recently approved by ITU-T Study Group 10, which will lead to a new Recommendation by 2003. Tool support, current research on UCMs, and related notations are also discussed.

1 Introduction

The modeling of reactive systems requires an early emphasis on behavioral aspects such as interactions between the system and the external world (including the users), on the cause-to-effect relationships among these interactions, and on intermediate activities performed by the system. Scenarios are particularly good at representing such aspects so that various stakeholders can understand them.

Owing to their distributed and critical nature, telecommunications systems are representative of complex reactive systems. Emerging telecommunications services and features require industries and standardization bodies (ANSI, ETSI, ISO, ITU, TIA, IETF, etc.) to describe and design increasingly complex functionalities, architectures, and protocols. This is especially true of wireless systems, where the mobility of users brings an additional dimension of complexity. Recent and upcoming technologies based on agents, XML, and IP, which involve complex and sometimes unpredictable policy-driven negotiations between communicating entities, also raise new modeling issues as protocols and entities become more dynamic in nature and evolve at run time.

The design and standardization of telecommunication systems and features results from a design process frequently comprised of three major stages, shown in Figure 1. At stage 1, features are first described from the user's point of view in prose form, with use cases, and with tables. The focus of the second

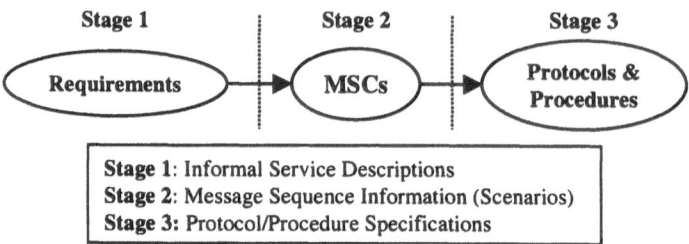

Fig. 1. Three-stage methodology

stage is on control flows between the different entities involved, represented using sequence diagrams or *Message Sequence Charts* (MSCs [22]). Finally, stage 3 aims to provide (informal) specifications of protocols and procedures. Formal specifications are sometimes provided (e.g. in SDL [21]), but overall they still suffer from a low penetration [10,17], especially in North-America [2,18]. ITU-T developed this three-stage methodology two decades ago to describe services and protocols for ISDN. Naturally, such descriptions emphasize the reactive and behavioral nature of telecommunications systems. In this methodology, scenarios are often used as a means to model system functionalities and interactions between the entities such that different stakeholders may understand their general intent as well as technical details.

Due to the inherent complexity and scale of emerging telecommunications features, special attention has to be brought to the early stages of the design process. The focus should be on system and functional views rather than on details belonging to a lower level of abstraction, or to later stages in this process. Many ITU-T members recognize the need to improve such three-stage process in order to cope with the new realities cited above. In particular, Study Group 10, which is responsible for the evolution of standards such as MSC, SDL, and TTCN, recently approved a question for study during the next ITU-T Study Period to develop a User Requirements Notation (URN) based on scenarios. The objective of the question is to develop a new Recommendation by the year 2003 [25].

This question focuses on what notation may be developed to complement MSCs, SDL and UML in capturing user requirements in the early stages of design when very little design detail is available. Such notation should be able to describe features as user requirement scenarios without any reference to states, messages or system components. Reusability of scenarios across a wide range of architectures is needed with allocation of scenario responsibilities to architectural components. The notation should enable simpler modeling of dynamic systems, early performance analysis at the requirements level, and early detection of undesirable interactions among features or scenarios.

While UML activity diagrams provide some capability in this area [27], a requirements notation with dynamic refinement capability and better allo-

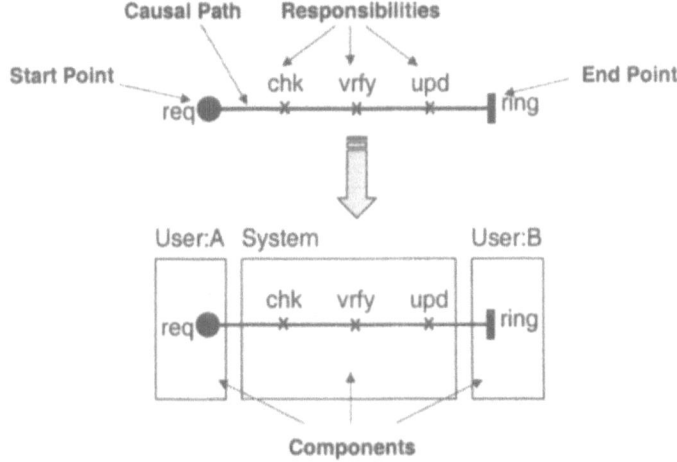

Fig. 2. Simple Use Case Map

cation of scenario responsibilities to architectural components is required [7]. The Use Case Map notation offers such capabilities. The basics of this notation and its main benefits are illustrated in Section 2. Section 3 presents various domains of application of UCMs as a feature description notation followed by an overview of current tool support, research directions, related notations, and formalization issues. A brief conclusion follows in Section 4.

2   Use Case Maps

2.1   Basic Notational Elements

The *Use Case Map* (UCM) notation [13,14] is used for describing *causal relationships* between *responsibilities*, which may potentially be bound to underlying organizational structures of abstract *components* (see Figure 2). Responsibilities are generic and can represent actions, activities, operations, tasks to perform, and so on. Components are also generic and can represent software entities (objects, processes, databases, servers, functional entities, network entities, etc.) as well as non-software entities (e.g. users, actors, processors). The relationships are said to be causal because they involve concurrency and partial orderings of activities and because they link causes (e.g., preconditions and triggering events) to effects (e.g. postconditions and resulting events). In a way, UCMs show related use cases in a map-like diagram.

The scenario in Figure 2 represents a simplified call connection initiated by user A on **req**. The system first checks whether the call should be allowed (**chk**) and then verifies whether the called party is busy or idle (**vrfy**). In both cases here, we assume that the call request goes through. Then, the system status is updated (**upd**) and a resulting ringing event occurs at B's side (**ring**).

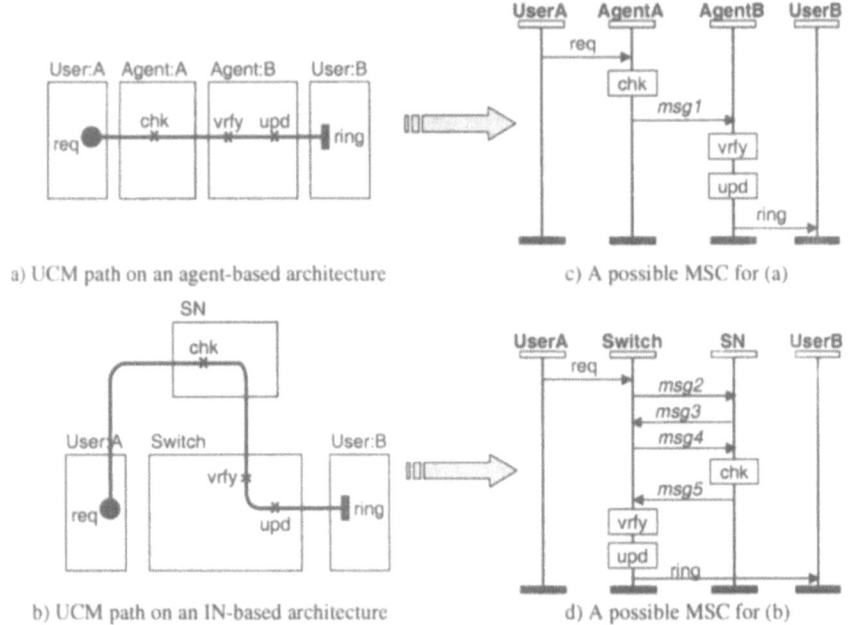

a) UCM path on an agent-based architecture   c) A possible MSC for (a)

b) UCM path on an IN-based architecture   d) A possible MSC for (b)

Fig. 3. UCM path bound to two different component structures, and potential MSCs

2.2   UCMs, Architectures and Messages

UCM paths and their responsibilities are useful for describing features at an early stage in the design cycle (e.g. at stage 1), even when no component is involved (e.g. Figure 5). It is then possible to bind UCM paths to a suitable structure of components, which leads to a visual integration of scenarios and architecture in a single view. UCM scenario paths possess a high degree of reusability and they lead to behavioral patterns that can be utilized across a wide range of applications. For instance, the UCM path from Figure 2 can be bound to alternative architectures therefore enabling early architectural reasoning. Figure 3(a) uses an agent-based architecture whereas Figure 3(b) uses a more conventional architecture based on Intelligent Networks (IN).

UCM paths are also more likely to survive evolutions and other modifications to the underlying architecture than scenarios described in terms of message exchanges or interactions between components. For instance, Figure 3(c) is an MSC capturing the scenario from Figure 3(a) in terms of message exchanges. This is a straightforward interpretation with artificial messages (in italic characters). In this system, each user can communicate with its agent only, and agents can communicate with other agents. Other such MSCs could possibly be derived from the same scenario path.

Figure 3(d) is a potential MSC extracted from the same scenario path, but this time bound to an IN-based architecture. Complex protocols or negotiation mechanisms could be involved between the switch and the service node, hence resulting in multiple messages. Communication constraints (not shown here) could also prevent users from communicating directly with service nodes; therefore the switch needs to be involved as a relay to refine the causal relationship between req and chk.

When extracting MSC-like scenarios from informal requirements, as it is often done in the three-stage methodology shown in Figure 1, many design decisions become buried in the details of the scenarios. For instance, the Wireless Intelligent Network (WIN) standard attempts to use the IN reference model but, due to legacy descriptions of former versions of the ANSI-41 North-American wireless standard, it only provides MSC scenarios where the components represent network elements belonging to the physical plane [9]. Design decisions such as the allocation to UCM responsibilities to functional entities (the logical components in the distributed functional plane) and the allocation of functional entities to network entities are lost [3]. Since this standard does not impose a specific mapping of functional entities to network entities, different vendors who build network entities may use different mappings (and this is actually happening). Designers must reverse-engineer information and scenarios that would be explicit in a UCM view where responsibilities and other constructs are bound to functional entities. This delays the design and implementation phases and leads to multiple interoperability problems. Such problems are unfortunately common in standards.

By using a UCM view, many issues related to messages, protocols, communication constraints, and structural evolutions (e.g. from one version of the structure to the next) can be abstracted from, and the focus can be put on intended functionalities and on reusable causal scenarios in their structural context.

2.3   UCMs and Scenario Integration

UCMs can also help structuring and integrating scenarios in various ways, e.g. sequentially, as alternatives (with OR-forks/joins) or concurrently (with AND-forks/joins). However, one of the most interesting constructs for scenario integration is certainly the *dynamic stub*, shown as a dashed diamond in Figure 4.

While static stubs (not shown here) contain only one sub-map (called *plug-in*), dynamic stubs may contain multiple sub-maps whose selection can be determined at run-time according to a *selection policy*. Such a policy can make use of preconditions, assertions, run-time information, composition operators, etc. in order to select the plug-in(s) to use. Selection policies are described with a (formal or informal) language suitable for the context where they are used. The plug-in maps are sub-maps that describe locally how a

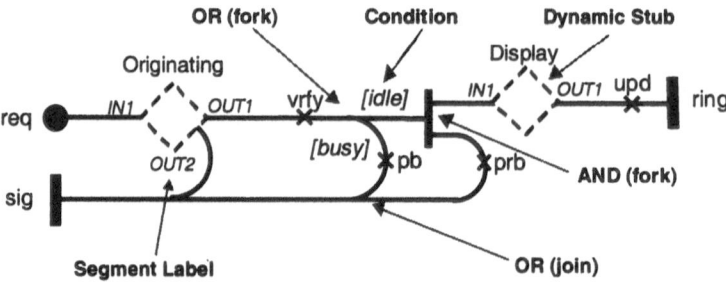

Fig. 4. Basic Call UCM with two dynamic stubs

feature modifies the basic behavior. Multiple levels of stubs and plug-ins can be used.

Figure 5 shows four UCMs. The top-level UCM is the Basic Call of Figure 4, which contains two dynamic stubs. Each of these stubs includes a DE-FAULT plug-in (which happens to be the same in both cases) that represents how the basic call reacts in the absence of other features.

The Originating stub has two plug-ins:

- *Originating Call Screening* (OCS), which checks whether the call should be denied or allowed (chk). When denied, an appropriate event is prepared (pd) and signaled (sig). Its *binding relationship*, which connects the input/output segments of a stub to the start/end points of its plug-in, is $\{\langle IN1, in1\rangle, \langle OUT1, out1\rangle, \langle OUT2, out2\rangle\}$.
- TEENLINE, which denies the call provided that the request is made during a specific time interval and that the personal identification number (PIN) provided is invalid or not entered in a timely manner. The zigzag path leaving the timer represents a timeout path. The binding relationship for this feature is also $\{\langle IN1, in1\rangle, \langle OUT1, out1\rangle, \langle OUT2, out2\rangle\}$.

The Display stub contains only one feature:

- *Call Number Delivery* (CND), which displays the number of the originating party (disp) concurrently with the rest of the scenario (update and ringing). The binding relationship is $\{\langle IN1, in1\rangle, \langle OUT1, out1\rangle\}$.

Adding features to such UCM collections is often achieved by creating new plug-ins for the existing stubs, or by adding new stubs containing either new plug-ins or instances of existing plug-ins. In all cases, the selection policies need to be updated appropriately.

Stubs and selection policies tend to localize the places on scenarios where undesirable feature interactions can occur [4,6,26], hence simplifying the analysis. They can also be used to specify priorities of some features over others. For instance, TEENLINE could be given a sequential priority over OCS in the Originating stub. Many spurious interactions between features are hence avoided by structuring and integrating the scenarios in the proper context.

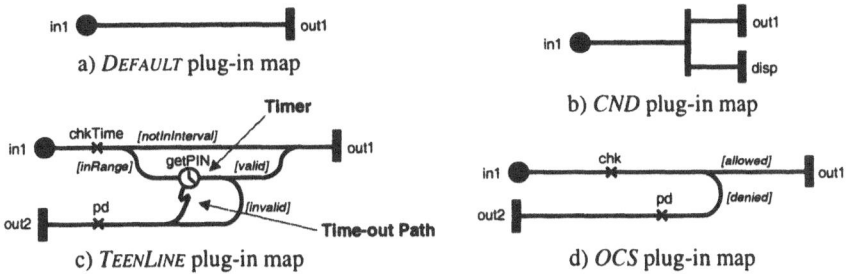

Fig. 5. Basic Call UCM with its four plug-ins

2.4 Performance and Agent-Oriented Annotations

The UCM core notation has been extended over the years to cover different fields of applications such as performance analysis at the requirements level (with *timestamps*) [30,31] and agent-oriented design (with *goal tags*). These areas are also recognized by ITU-T Study Group 10 as relevant to the description of features.

Timestamps are located on UCM paths in order to identify various performance constraints and response time requirements. For example, Figure 6 shows two timestamps attached to a UCM. Response time requirements (expected response time, probability, etc.) can be defined between pairs of timestamps (e.g., T1 and T2) at the scenario path level. In order to generate useful performance simulation models, other notation elements can be annotated, for instance start points with arrival characteristics (exponential, deterministic, uniform, Erlang, etc.) and responsibilities (associated data store, performed service requests, etc.).

Goal tags are used to associate high-level goals and intentions to UCM scenario paths. This is particularly relevant to the description of agent systems, where distributed goals are intended to be achieved by collaborating agents. Figure 7 shows the symbol used to represent goal tags. Similar to

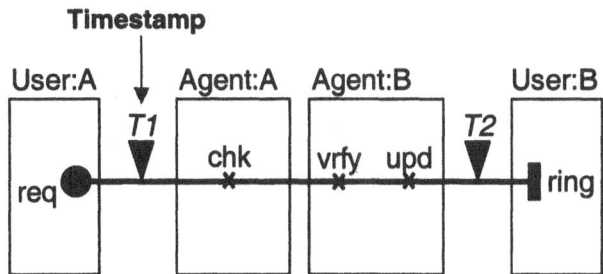

Fig. 6. Performance annotations with timestamps

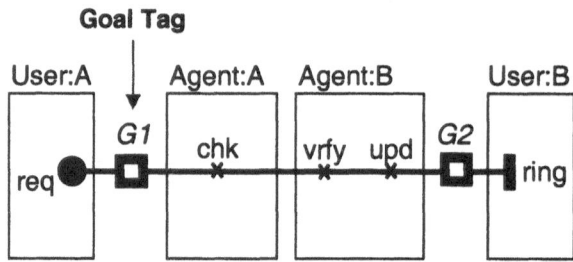

Fig. 7. Agent-oriented goal annotations with goal tags

timestamps, these tags can be coupled in pairs to describe goals and their pre/post-conditions.

2.5 Additional Notation Elements

The UCM notation contains many additional elements, including:

- *Component types and attributes:* types can be used to visually distinguish active components (e.g. processes), passive components, containers, agents, etc. A component may also possess several attributes (e.g. replication factor and mutual exclusion).
- *Dynamic components and dynamic responsibilities:* dynamic stubs show how behavior patterns can evolve at run time. A structure of component can also evolve at run time through the use of dynamic components, which represent, in a static way, roles that can be filled by actual instances of components at different times. Dynamic responsibilities are used to create, delete, store, retrieve, or move dynamic components.
- *Path interactions:* UCM paths can also be combined through the use of synchronous or asynchronous path interactions, which often involve a special point on a path called *waiting place.*
- *Aborts:* a scenario path can abort the evolution of another path through this construct. This can also be used to model exceptions.
- *Failure points:* failure points represent explicit places on a UCM path where a scenario could stop before reaching its end point. This is useful when robustness needs to be addressed at a high level of abstraction.

Although these advanced constructs are often useful for the description of features, they will not be discussed further in this paper. The interested reader can however consult the UCM Quick Reference Guide (Appendix A) and the UCM virtual library of the *UCM User Group* for more information [34].

3   UCMs as a Feature Description Notation

Use Case Maps have a number of properties that satisfy many of the requirements described in Section 1: scenarios can be mapped to different architectures, variations of run-time behavior and structures can be expressed, and scenarios can be structured and integrated incrementally in a way that facilitates the early detection of undesirable interactions and the early evaluation of performance. Performance becomes a property of paths, rather than just a non-functional property of a whole system, as it is often considered to be [14,30,31]. Macroscopic behavior patterns are described independently of details belonging to connections between components (e.g. messages), and large-scale dynamic situations and issues can be made visible at a glance.

The UCM notation is currently applicable to a wide range of areas, it is already supported by a tool, and it is the topic of several research projects. It also possess interesting characteristics that are difficult to find all at once in other notations for describing features. This will be discussed briefly in the following sections.

3.1   Areas of Application

Use Case Maps are well suited for describing requirements and high-level designs of reactive (event-driven) and distributed systems and their features. UCMs have a history of applications to the description of features and telecommunications systems of different natures (e.g. [3–6,8]), to the avoidance and detection of undesirable interactions between scenarios or features (e.g. [4,6,15,26]) and to early performance analysis (e.g. [30,31]).

UCMs are however not restricted to telecommunications systems. They are also being used to describe systems and features from various domains such as (in no particular order) airline reservation applications, elevators, railway control systems, agent systems, network management applications, Web applications, graphical user interfaces, drawing packages, multimedia applications, banking applications, object-oriented frameworks, "work patterns" of software engineers, and many others [34].

3.2   UCM Navigator Tool

The UCM notation is also supported by a tool: the *UCM Navigator* [24]. This tool has been developed at Carleton University (Ottawa) in order to help drawing correct UCMs. Among other features, this tool supports the path and component notations found in Appendix A, and it maintains various bindings (plug-ins to stubs, responsibilities to components, sub-components to components, etc.). Also, it allows users to navigate much like a Web browser, and to visit and edit the plug-ins related to stubs of all levels (Figure 8).

The UCM Navigator is transformation-based and it ensures that UCMs are syntactically correct by construction. The tool saves, loads, exports and

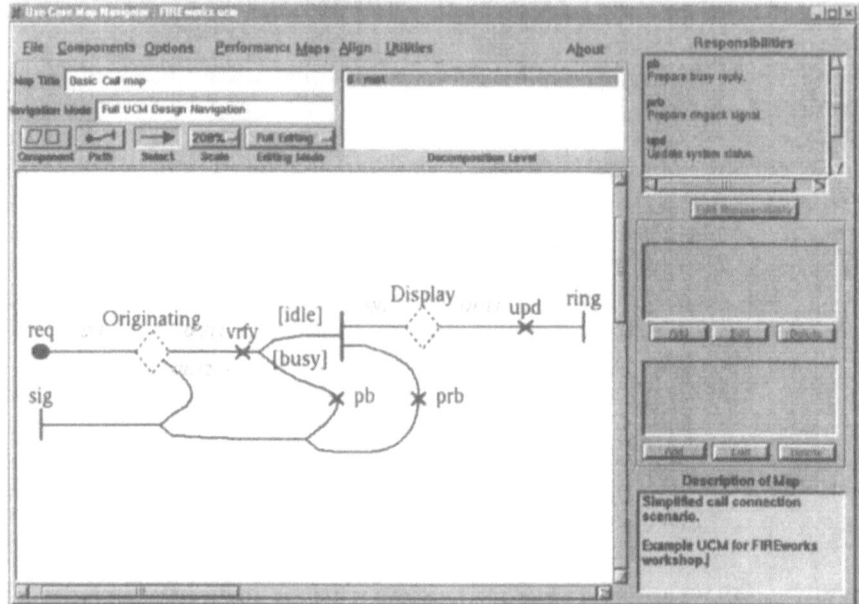

Fig. 8. UCM Navigator screen shot

imports UCM as XML files, which are valid according to a UCM Document Type Definition (DTD) [34]. This DTD describes the current formal definition of UCMs, which is based on hypergraphs, and the UCM Navigator ensures that static semantic rules and other constraints are satisfied. No formal dynamic semantics currently exists for the UCM notation. The tool can also export UCM figures in three formats: Encapsulated PostScript (EPS), Maker Interchange Format (MIF), and Computer Graphics Metafile (CGM). Flexible reports can be generated as PostScript files ready to be transformed into hyperlinked and indexed PDF files. Multiple platforms are currently supported: Solaris, Linux (Intel and Sparc), HP/UX, and Windows (95, 98, 2000 and NT).

3.3   Current UCM Research

Many researchers are developing links between Use Case Maps and other notations and languages, something that is necessary in order to produce concrete implementations and products from features and other functional requirements captured as UCMs. Among others, UCMs are currently being mapped to:

- (H)MSCs or UML sequence diagrams (e.g. Figure 3) [6,12]. This generation is in the process of being formalized and automated in the UCM Navigator tool.

- Hierarchical state machines such as UML statechart diagrams or ROOM-charts [12].
- LOTOS models, which enable formal validation, verification, and detection of undesirable interactions [4–6,8].
- SDL models, which also enable formal validation [29].
- UML and UML-RT models [7,12].
- Other research projects include the generation of performance models (e.g. in Layered Queuing Networks-LQNs), of abstract test cases for functional testing (e.g. in the Tree and Tabular Combined Notation-TTCN), and of programs in agent-oriented languages.

The notation is also evolving under the guidance of a newly created *UCM Working Group* composed of members from industry and universities. In particular, this group intends to present a contribution to ITU-T proposing Use Case Maps as an appropriate notation to capture *functional* requirements (a companion notation such as the one in the NFR framework [16] would have to capture non-functional requirements). UCMs would hence represent part of the answer to the User Requirements Notation (URN) question.

The multiple connections between UCMs and other languages enable the creation of many design trajectories relevant to telecommunications systems, as suggested in the introduction. In particular, we envision the following trajectory, inspired from [2,3,12,18]: requirements capture and architectural reasoning is done with UCM/URN (stage 1), which are first transformed into MSCs or interaction diagrams (stage 2), then into state machines in SDL or UML-RT statechart diagrams (stage 3), and finally into concrete implementations (possibly through semi-automated code generation). Inspection, validation, verification, performance analysis, interaction detection, and test generation can be performed to various degrees at all stages.

3.4 UCMs and Other Notations for Describing Features

Features can be described in a number of ways, for instance with goals, logical properties, and scenarios. These approaches are not necessarily mutually exclusive (e.g. goals can be associated to UCM scenarios, as seen in Section 2.4). Goals and properties are usually large-grained, declarative, and cover more situations than scenarios, which are more operational and which capture partial and non-exhaustive views of the system. However, the discovery and structuring of goals and properties is not an easy task, whereas the construction of scenarios is often simpler [28]. Scenarios are also more in line with current practices in ITU-T and other standardization bodies. According to many people in such organizations, it is better to find a practical notation that will improve the current situation and that will be used than to aim for the best theoretical solution, which is unlikely to be used at all [18].

Scenarios can also be textual rather than graphical (e.g. Jacobson's use cases [23]). Although textual scenarios can adequately describe requirements in many situations, especially when appropriate construction guidelines are provided [11], these scenarios are still very much linear in nature. Graphical notations such as UCMs enable the description of scenarios in two dimensions as well as the compact representation of multiple scenarios.

Still, many graphical scenario notations could be considered as good candidates for URN in the context of feature descriptions. Here is a brief comparison between UCMs and four such notations, namely MSCs, Chisel diagrams, Petri Nets, and UML activity diagrams:

- *Message Sequence Charts (MSCs)*: as shown in Section 2, MSCs and similar interaction diagrams suffer from a premature commitment to messages and components. This is not always appropriate for the early stages of feature definitions because details irrelevant to the requirements level must be considered, ·and this in turn hides the original intent of the feature. UCMs abstract from messages and improve the reusability of scenarios across component architectures.

- *Chisel diagrams*: Aho et al. have performed empirical studies with telecommunication engineers to create the Chisel notation [1]. Chisel diagrams are trees whose branches represent sequences of (synchronous) events taking place on component interfaces. Nodes describe these events (multiple concurrent events can take place in one node) and arcs, which can be guarded by conditions, link the events in causal sequences. Chisel diagrams describe multiple abstract scenarios, but like MSCs they focus on inter-component interactions even if the components are hidden from this particular view. UCMs can abstract from these events, and UCMs also support advanced concepts such as dynamic stubs, which have no equivalent in the Chisel notation. Note that similar to the UCM-LOTOS translation mentioned in Section 3.3, a Chisel-LOTOS mapping has been defined and automated by Turner [33].

- *Petri Nets (PNs)*: these are abstract machines used to describe system behaviour visually with a directed graph containing two types of nodes: places and transitions. Basic PNs suffer from a state explosion problem when complex problems are addressed. Various extensions for data types and modules, which help to cope with this problem, are currently being standardized as High-Level Petri Nets [20]. The main benefits of PNs over UCMs are their formality and executability. However, their use as a requirements notation for features (if such thing exists) is still remote from the current ITU-T practice, and PNs also lack concepts such as dynamic stubs and a view that combines visually behavioral scenarios and component architectures. PNs could however represent a candidate language for formalizing the UCM dynamic semantics.

- *UML activity diagrams*: probably the next best alternative to UCMs for URN, activity diagrams share many characteristics with UCMs [27]: focus on sequences of actions, guarded alternatives, and concurrency; start and end points from each notation have a similar purpose; and complex activities can be refined. However, activity diagrams are usually unrelated to components (*swimlanes*, which could be seen as enabling bindings of activities to components, provide functional grouping only), and their sub-diagrams are less flexible and less powerful than UCM stubs. Activity diagrams could however be evolved to support these attractive UCM concepts [7], which would then help satisfy ITU-T's goal of linking URN to MSC, SDL, and UML.

3.5   On the Formalization of UCMs

The UCM notation enjoys an enthusiastic community of users and it has been used successfully as a feature description notation in the domains of telecommunications and other reactive systems. Several users however complain about the lack of formal foundations behind the notation, and work is being done to address this situation (e.g. [7]). We like to think of UCMs as a back-of-envelope notation that offers an attractive level of abstraction for describing feature requirements at early stages, without getting dragged into low-level details. An unfortunate consequence is that UCMs are open to misinterpretation as much as any other non-formal notation. Conventions, standard styles, and patterns can help to cope with this issue to some extent, but formalization is required as well. However, finding the appropriate degree of formalization, which would not sacrifice the appealing and semi-formal characteristics of UCMs, remains an issue. The URN effort represents a great opportunity for finding a practical solution to this problem. Formalization could also be done by integrating UCMs to UML, for instance by adapting and improving the activity diagrams notation and semantics. High-level Petri Nets could also be used as a semantic model. In any case, UCMs possess several attractive constructs and concepts that should be seriously considered in any full-fledged feature description language.

4   Conclusion

Scenarios are a popular and practical approach to the design of reactive systems. The UCM notation enables the early description of features in terms of causal scenarios bound to underlying abstract components. This paper illustrates the core constructs of the notation, its main benefits, and its place in common design trajectories. It also shows that the UCM notation satisfies many of the needs expressed in ITU-T's question for study about a User Requirements Notation.

The UCM notation is evolving and is the target of many research and development projects. Industrial partners appreciate this notation because it allows for senior designers, system architects and product managers, who possess good knowledge of the domain and of existing systems, to "work again", i.e. they can communicate their knowledge efficiently to junior designers, who then take care of the specifics. Senior people do not need to know all the details surrounding recent or emerging technologies to describe desired features in a way that design teams can understand and refine. We are hence very confident in the usefulness of the notation and of the level of abstraction it addresses.

Acknowledgements

This work results from collaborations and discussions with many colleagues, co-authors, and partners from the University of Ottawa's LOTOS group, the UCM User Group, Mitel Corporation and Nortel Networks. I am also indebted towards my supervisor, Luigi Logrippo, and my former co-supervisor, Ray Buhr, for their encouragement. Finally, I would like to thank Communications and Information Technology Ontario (CITO), the Natural Sciences and Engineering Research Council of Canada (NSERC), Mitel Corporation and Nortel Networks for their financial support.

References

1. Aho, A., Gallagher, S., Griffeth, N., Scheel, C. and Swayne, D.: "Sculptor with Chisel: Requirements Engineering for Communications Services". In: *Fifth International Workshop on Feature Interactions in Telecommunications and Software Systems (FIW'98)*, Lund, Sweden, October 1998. IOS Press, Amsterdam, 45–63.
2. Amyot, D., Andrade, R., Logrippo, L., Sincennes, J., and Yi, Z.: "Formal methods for mobility standards". In: *Proc. of the 1999 IEEE Emerging Technologies Symposium on Wireless Communications and Systems*, Richardson, Texas, USA, April 1999.
3. Amyot, D. and Andrade, R.: "Description of Wireless Intelligent Network Services with Use Case Maps". In: *SBRC'99, 17th Brazilian Symposium on Computer Networks*, Salvador, Brazil, May 1999.
4. Amyot, D., Buhr, R. J. A., Gray, T., and Logrippo, L.: "Use Case Maps for the Capture and Validation of Distributed Systems Requirements". In: *RE'99, Fourth IEEE International Symposium on Requirements Engineering*, Limerick, Ireland, June 1999, 44–53.
5. Amyot, D. and Logrippo, L.: "Use Case Maps and LOTOS for the Prototyping and Validation of a Mobile Group Call System". In: *Computer Communication*, 23(12), 1135–1157, May 2000.
6. Amyot, D., Charfi, L., Gorse, N., Gray, T., Logrippo, L., Sincennes, J., Stepien, B. and Ware, T.: "Feature Description and Feature Interaction Analysis with Use Case Maps and LOTOS". In: *Sixth International Workshop on Feature Interactions in Telecommunications and Software Systems (FIW'00)*, Glasgow, Scotland, UK, May 2000.
7. Amyot, D. and Mussbacher, G.: "On the Extension of UML with Use Case Maps Concepts". In: *<<UML>>2000, 3rd International Conference on the Unified Modeling Language*, York, UK, October 2000.

8. Andrade, R.: "Applying Use Case Maps and Formal Methods to the Development of Wireless Mobile ATM Networks". In: *Lfm2000: The Fifth NASA Langley Formal Methods Workshop*, Williamsburg, Virginia, USA, June 2000.

9. ANSI/TIA/EIA: *ANSI 771, Wireless Intelligent Networks (WIN). Additions and modifications to ANSI-41 (Phase 1)*. TR-45.2.2.4, December 1998.

10. Ardis, M. A., Chaves, J. A., Jagadeesan, L. J., Mataga, P., Puchol, C., Staskauskas, M. G., and Olnhausen, J. V.: "A Framework for Evaluating Specification Methods for Reactive Systems - Experience Report". In: *Transactions on Software Engineering*, IEEE, 22 (6), 1996, 378–389.

11. Ben Achour, C., Rolland, C. Maiden, N. A. M. and Souveyet, C.: "Guiding Use Case Authoring: Results of an Empirical Study". In: *RE'99, Fourth IEEE International Symposium on Requirements Engineering*, Limerick, Ireland, June 1999, 36–43.

12. Bordeleau, F.: *A Systematic and Traceable Progression from Scenario Models to Communicating Hierarchical Finite State Machines*. Ph.D. thesis, SCS, Carleton University, Ottawa, Canada, August 1999.

13. Buhr, R. J. A. and Casselman, R. S.: *Use Case Maps for Object-Oriented Systems*, Prentice-Hall, 1996.

14. Buhr, R. J. A.: "Use Case Maps as Architectural Entities for Complex Systems". In: *Transactions on Software Engineering*, IEEE, December 1998, 1131–1155.

15. Buhr, R. J. A., Amyot, D., Elammari, M., Quesnel, D., Gray, T., and Mankovski, S.: "Feature-Interaction Visualization and Resolution in an Agent Environment". In: *Fifth International Workshop on Feature Interactions in Telecommunications and Software Systems (FIW'98)*, Lund, Sweden, October 1998. IOS Press, Amsterdam, 135–149.

16. Chung, L., Nixon, B. A., Yu, E. and Mylopoulos, J.: *Non-Functional Requirements in Software Engineering*. Kluwer Academic Publishers, 2000.

17. Craigen, D., Gerhart, S., and Ralston, T.: *Industrial applications of formal methods to model, design, and analyze computer systems: an international survey*. Noyes Data Corporation (Publisher), USA, 1994.

18. Hodges, J. and Visser, J.: "Accelerating Wireless Intelligent Network Standards Through Formal Techniques". In: *IEEE 1999 Vehicular Technology Conference (VTC99)*, Houston, USA, 1999.

19. ISO, Information Processing Systems, OSI: LOTOS - *A Formal Description Technique Based on the Temporal Ordering of Observational Behaviour*. IS 8807, Geneva, Switzerland, 1989.

20. ISO/IEC: *High Level Petri Net Standard*, DIS 15909, JTC 1/SC 7, Geneva, Switzerland, 1999.

21. ITU-T: *Recommendation Z.100, Specification and Description Language (SDL)*. Geneva, Switzerland, 2000.

22. ITU-T: *Recommendation Z.120, Message Sequence Chart (MSC)*. Geneva, Switzerland, 2000.

23. Jacobson, I.: "The Use Case Construct in Object-Oriented Software Engineering". In: John M. Carroll (ed.), *Scenario-Based Design: Envisioning Work and Technology in System Development*. John Wiley and Sons, 1995, 309–336.

24. Miga, A.: *Application of Use Case Maps to System Design with Tool Support*. M.Eng. thesis, Dept. of Systems and Computer Engineering, Carleton University, Ottawa, Canada, October 1998.

25. Monkewich, O.: *New Question 12: URN: User Requirements Notation.* Canadian contribution to ITU-T Study Group 10, COM10-D56, November 1999.
26. Nakamura, M., Kikuno, T., Hassine, J., and Logrippo, L.: "Feature Interaction Filtering with Use Case Maps at Requirements Stage". In: *Sixth International Workshop on Feature Interactions in Telecommunications and Software Systems (FIW'00)*, Glasgow, Scotland, UK, May 2000.
27. Object Management Group: *Unified Modeling Language Specification, Version 1.3.* June 1999.
28. Rolland, C., Souveyet, C. and Ben Achour, C.: "Guiding Goal Modelling using Scenarios". In: *IEEE Transactions on Software Engineering, Special Issue on Scenario Management.* Vol. 24, No. 12, December 1998.
29. Sales, I. and Probert, R. L.: "From High-Level Behaviour to High-Level Design: Use Case Maps to Specification and Description Language". In: *SBRC'2000, 18th Brazilian Symposium on Computer Networks*, Belo Horizonte, Brazil, May 2000.
30. Scratchley, W. C. and Woodside, C. M.: "Evaluating Concurrency Options in Software Specifications". In: *MASCOTS'99, Seventh International Symposium on Modelling, Analysis and Simulation of Computer and Telecommunication Systems*, College Park, MD, USA, October 1999, 330–338.
31. Scratchley, W. C.: *Evaluation and Diagnosis of Concurrency Architectures.* Ph.D. thesis, Dept. of Systems and Computer Engineering, Carleton University, Ottawa, Canada, June 2000.
32. Selic, B.: "Turning Clockwise: Using UML in the Real-Time Domain". In: *Communications of the ACM*, 42(10), October 1999, 46–54.
33. Turner, K. J.: "Formalising the Chisel Feature Notation". In: *Sixth International Workshop on Feature Interactions in Telecommunications and Software Systems (FIW'00)*, Glasgow, Scotland, UK, May 2000. IOS Press, Amsterdam, 241–256.
34. *Use Case Maps Web Page and UCM User Group*, March 1999.
    http://www.UseCaseMaps.org

N.B. Many of these papers are available in the UCM virtual library:
http://www.UseCaseMaps.org/pub/

A   UCM Quick Reference Guide

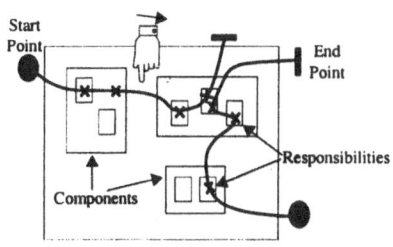

Imagine tracing a path through a system of objects to explain a causal sequence, leaving behind a visual signature. Use Case Maps capture such sequences. They are composed of:

- **start points** (filled circles representing preconditions and/or triggering causes)
- causal chains of **responsibilities** (crosses, representing actions, tasks, or functions to be performed)
- and **end points** (bars representing postconditions and/or resulting effects).

The responsibilities can be bound to **components**, which are the entities or objects composing the system.

A1. Basic notation and interpretation

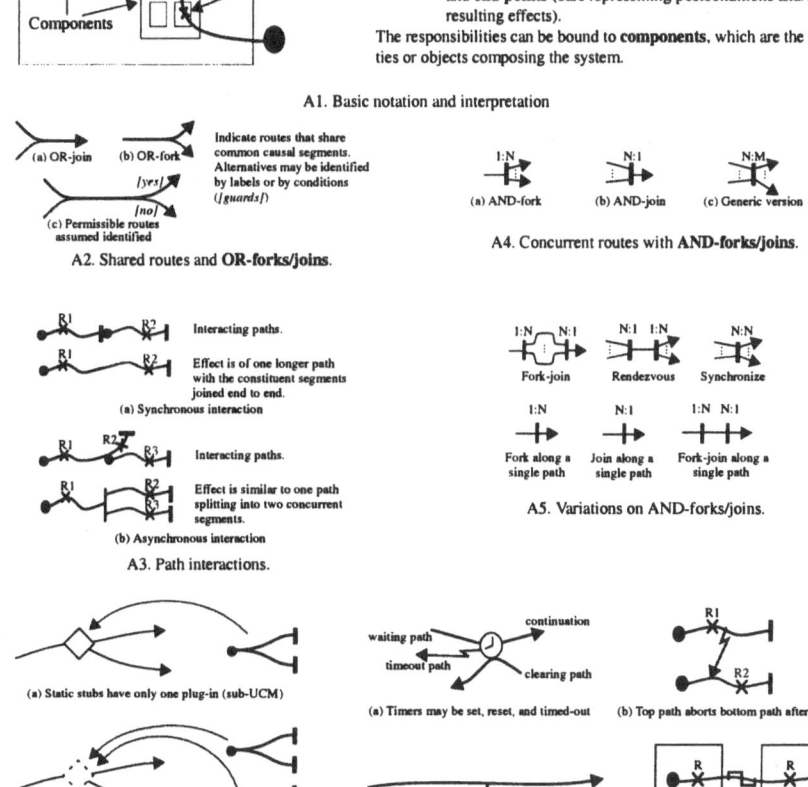

A2. Shared routes and **OR-forks/joins**.

A3. Path interactions.

A4. Concurrent routes with **AND-forks/joins**.

A5. Variations on AND-forks/joins.

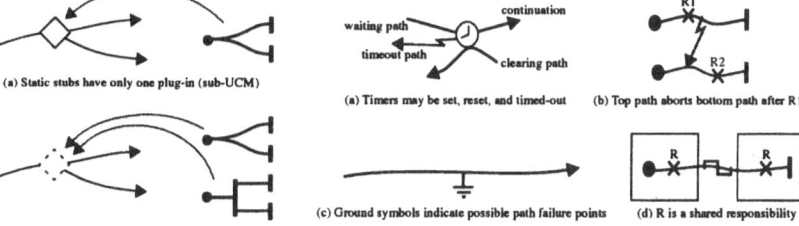

A6. Stubs and **plug-ins**.

A7. Timers, aborts, failures, and shared responsibilities.

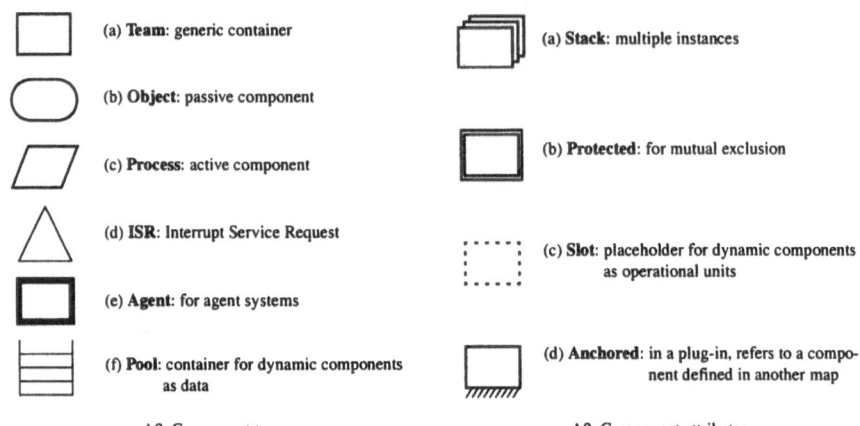

(a) **Team**: generic container

(b) **Object**: passive component

(c) **Process**: active component

(d) **ISR**: Interrupt Service Request

(e) **Agent**: for agent systems

(f) **Pool**: container for dynamic components as data

A8. Component types.

(a) **Stack**: multiple instances

(b) **Protected**: for mutual exclusion

(c) **Slot**: placeholder for dynamic components as operational units

(d) **Anchored**: in a plug-in, refers to a component defined in another map

A9. Component attributes.

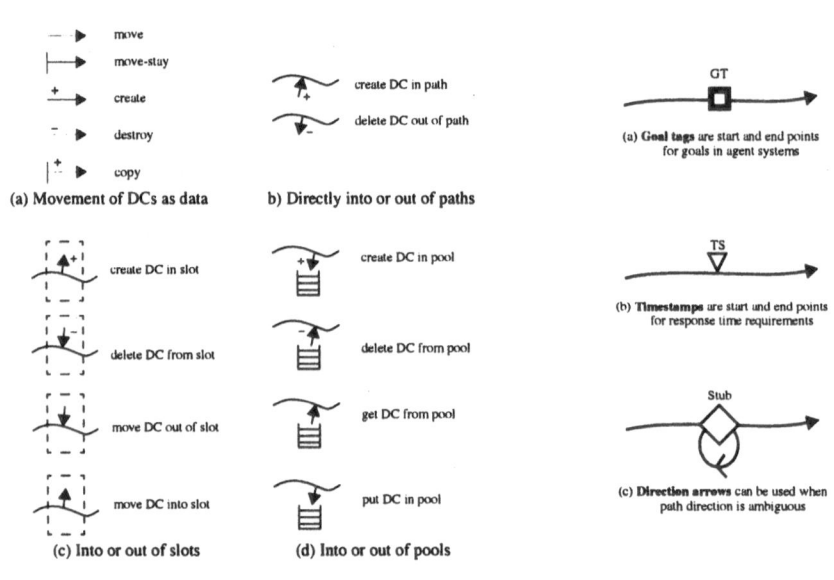

move

move-stay

create

destroy

copy

(a) Movement of DCs as data

create DC in path

delete DC out of path

b) Directly into or out of paths

create DC in slot

delete DC from slot

move DC out of slot

move DC into slot

(c) Into or out of slots

create DC in pool

delete DC from pool

get DC from pool

put DC in pool

(d) Into or out of pools

GT

(a) **Goal tags** are start and end points for goals in agent systems

TS

(b) **Timestamps** are start and end points for response time requirements

Stub

(c) **Direction arrows** can be used when path direction is ambiguous

A10. Movement notation for **dynamic components** (DCs).

A11. Notation extensions

An incremental method for the design of feature-oriented systems

Karim Berkani[1], Pascale Le Gall[2], and Francis Klay[1]

[1] France Télécom R&D, site de Lannion, Technopole Anticipa,
2 av. Pierre Marzin, F-22307 Lannion Cedex, France
[2] Université d'Evry, Laboratoire des Méthodes Informatiques, Bd des Coquibus,
F-91025 Evry Cedex, France

Abstract. Telecommunication systems involve many optional features which cannot be combined according to an obvious definition because they can interact together in an unpredictable way. In order to find the best way to combine features, a major difficulty is to foresee or to discover when a feature can disrupt an another one. Once such an interaction has been analysed, it still remains to define the best combination between the involved features. As this decision is purely subjective, we model the design of feature-oriented systems as an iterative process controlled by an expert judgement: all along the process, the expert increases his knowledge about the system under design and thus, may revise the feature integration design w.r.t. his expertise ... until he decides to accept the current feature integration as conform to his ideal view of the system. We propose to jointly adjust the description of a system built by integrating features and the expected properties on the global system all along the iterative process. Thus, step by step, the expert is brought to classify properties as desirable or undesirable.

1 Introduction

In the field of new telecommunication systems, systems are becoming really hard to design in a systematic way. Indeed, they involve many optional features which cannot be combined according to an obvious definition. Precisely, the feature interaction problem is a problem of integration. The integration of a feature on a system can modify the behaviour of the previously integrated features. A classical example of this phenomenon is the combination of the features known as respectively the Originating Called Screening one (OCS) and the Call Forward Unconditionally one (CFU). An interaction that one can observe in this case is that provided that the subscriptions to the feature are done according to a given schema (the phone B subscribes to the CFU feature with C as redirection while the phone C subscribes to OCS with the phone A in its screening list), then according to the way the two features have been combined, when A tries to call to B, a communication from A to C may be established by calling on the CFU feature even if this communication is forbidden by the OCS feature subscribed by C. The feature interaction problem consists first in discovering such interactions, then in choosing between accepting or rejecting these interactions which result from the joint integration of several features. The design of a system built as the integration of several features on a basic system starts from the descriptions of both the underlying basic system and the feature functionalities. These descriptions as well as the way the features are integrated may be done at any abstraction level (e.g. as an abstract specification or a precise implementation). We will tackle the feature interaction problem at a high level of abstraction. Thus, feature interactions, at least the ones which may considered as logical interactions, may be found as soon as possible during the design cycle. But even if we have formal descriptions for the basic system, the features and the way the features are integrated together, it remains a fundamental question: does the resulting system verify the expected properties ? Indeed, the global system defines a new behaviour corresponding to some compromises between the different requirements involved by the concerned features. Nothing guarantees that finally this global system is suitable with respect to the (informal) needs. Since 1990, industrial and academic organisations have engaged significant works (e.g. [3,4,6]) to bring solutions to the feature interaction problem. Different approaches have been studied. More or less, they all start from a description of the features within their a given formalism and then try to apply some verification techniques on their description. We are particularly interested in the way these works combine feature integration with interaction detection capacity. We can recognise three classes:

- Some work [13,5,22] introduce several ways to integrate features without taking into account of the verification part: features are integrated such that foreseeable interactions are kept or deleted depending on whether designers classify them as desirable or not. Main efforts are set on the definition of specialised feature composition operators allowing to prevent the appearance of undesirable feature interactions.
- Most work (e.g. [7], [9]) focus on the detection of interactions for feature-oriented systems where features are all integrated in a systematic way. In other words, a general and uniform composition operator is applied to model the integration of features. Then, in order to find feature interactions, the resulting system is in general checked against the expected properties associated to the integrated features. In these approaches, what does really matter is the capacity to compare a given description of the feature-oriented system with respect to the reference properties.
- Some other work (e.g. [12,19]) are based on the use of several composition constructs and on some verification techniques to detect interactions.

Most of these works involve effective verification methods assisted by tools: [16] with the proof system Spike, [19] with the model checker SMV and [9] with the Lutess tool. Indeed, the size and complexity of the telecommunication system descriptions make clear the need of tools for helping the designer. With this difference that we want to give more flexibility when designing feature-oriented systems by giving a most important part to expert decisions, we share numerous common points with all these mentioned works. In two recent contests [18,15], it has been noticed that the mentioned interactions partly depend on the point of view followed by the different teams. This lack of precise characterisations for desirable and undesirable interactions is due to the fact that the feature interaction problem introduces subjectivity. Indeed, typically, the integration of features on a given system provides a new description. Then, the central problem is to know what is expected from this new description. So, this adequation of feature integration with respect to the informal needs cannot be fully automatically decided. So, we propose to integrate features on the base of expert decisions. In this article, we develop a first proposal toward the integration of features guiding by expert decisions. More or less, feature integration is incrementally defined according to the interactions discovered and classified by the expert as desirable or not.

In section 2, we introduce the first generic elements for modelling the design of feature-oriented systems. In section 3, we explain how a methodology based on expert decisions allows to incrementally design feature-oriented systems. In section 4, we briefly describe a particular formalism provided with some dedicated composition operators useful to integrate features. In section 5, we illustrate our approach with a small example.

2   Specifications of feature-oriented systems

2.1   Notations

We first define some general notions or notations useful to introduce our incremental method for specifying feature-oriented systems. They are given in a rather flexible way which embeds various classical approaches of formal specifications. It presents the advantage of allowing to discuss about methodology independently of the underlying formalism. Thus, signatures will be used to denote system interfaces, sentences will characterise the properties we want to express about systems under consideration, ... With respect to the framework of institutions defined in [14], our presentation is in fact both a simplified version since for example, in our case, the class of signatures is not provided with morphisms and a slight modification since we put a special emphasis on the notion of signature inclusions[1] in order to be able to study a specification resulting from the integration of several features with respect to the properties required by each feature being integrated in the system. These last properties are of course expressed on the signatures of the features, which are often simple subsignatures of the global signature of the specification.

Definition 1
A logic $\mathcal{L}$ is given by a syntax and a semantics.

---

[1] A interested reader can find in [8] a characterisation of the notion of signature inclusions in term of category theory and in [11] the use if this characterisation for specifying feature-oriented systems.

- *The syntax is defined by a notion of signatures. With each signature Σ, is associated a set of sentences $Sen(\Sigma)$. The class of signatures is denoted Sig and is provided with an order relation $\subseteq$ (i.e. reflexive, antisymmetric and transitive) and with two binary laws $\cup$ and $\cap$ such that for any pair of signatures (Σ_1, Σ_2) in Sig, $\Sigma_1 \cap \Sigma_2$ (resp. $\Sigma_1 \cup \Sigma_2$) is the unique maximal (resp. minimal) element Σ of Sig verifying $\Sigma \subseteq \Sigma_1$ and $\Sigma \subseteq \Sigma_2$ (resp. $\Sigma_1 \subseteq \Sigma$ and $\Sigma_2 \subseteq \Sigma$). Moreover, we impose that for any Σ_1, Σ_2 of Sig satisfying $\Sigma_1 \subseteq \Sigma_2$, then $Sen(\Sigma_1) \subseteq Sen(\Sigma_2)$.*
- *The semantics is defined for each signature Σ of Sig by a class of Σ-models, denoted $Mod(\Sigma)$, and by a satisfaction relation $\models_\Sigma \subseteq Mod(\Sigma) \times Sen(\Sigma)$. Given φ a Σ-sentence and M a Σ-model, we denote as usual $M \models_\Sigma \varphi$ the fact that (M, φ) belongs to the relation $\models_\Sigma$. We extend the notation as follows: for a set Γ of Σ-sentences and a set $\mathcal{M}$ of Σ-models, $M \models_\Sigma \Gamma$ (resp. $\mathcal{M} \models_\Sigma \Gamma$) means that $\forall \varphi \in \Gamma, M \models_\Sigma \varphi$ (resp. $\forall M \in \mathcal{M}, M \models_\Sigma \Gamma$). Moreover, when there is no ambiguity, the index Σ may be omitted: the index corresponds to the signature of the model.*

In the sequel, we suppose that a logic $\mathcal{L}$ is given according to the notations of the definition 1. According to these notations, we can now define what are for us specifications and features.

Definitions 2
- *A specification SP is defined by its signature $Sig(SP)$ (and we say that SP is a $Sig(SP)$-specification) in Sig and its class $Mod(SP)$ of $Sig(SP)$-models (i.e. $Mod(SP) \subseteq Mod(Sig(SP))$).*
  A specification SP is said to be axiomatic if there exists a set Ax of Σ-sentences such that the set of models of SP is equal to the set of models satisfying Ax: $Mod(SP) = \{M \in Mod(\Sigma) \mid M \models Ax\}$. Then, we can simply denote SP by the couple (Σ, Ax).
  The set of specifications is denoted $\mathcal{S}$.
- *A feature F is defined by its signature $Sig(F)$ in Sig (and we say that F is $Sig(F)$-feature) and by its set of specific properties $Spe(F)$ as a set of $Sig(F)$-sentences (i.e $Spe(F) \subseteq Sen(Sig(F)))$.*
  The set of features is denoted $\mathcal{F}$.
- *Let F be a Σ_1-feature and let SP a Σ_2-specification with $\Sigma_1 \subseteq \Sigma_2$, SP satisfies the feature F, denoted by $SP \models_{\Sigma_2} F$, if $Mod(SP) \models_{\Sigma_2} Spe(F)$.*
- *A composition operator is a symbol op provided with a natural number n as arity. Such a composition operator is interpreted as a partial function $\overline{op} : (\mathcal{F} \cup \mathcal{S})^n \to \mathcal{S}$ with $(\mathcal{F} \cup \mathcal{S})^n$ as domain and $\mathcal{S}$ as codomain. By denoting Θ the set of all composition operators, $\Theta(\mathcal{F})$ denotes the set of all well-formed terms only built from composition operators of Θ applying on features of $\mathcal{F}$. Given a term SP of $\Theta(\mathcal{F})$, we denote $Feat(SP)$ all the features of $\mathcal{F}$ occurring in the term SP and we say that SP results from the integration of the features of $Feat(SP)$ and when it exists, $\overline{SP}$ denote the interpretation of SP obtaining by composing the different interpretations of symbols occurring in SP.*

- Let SP be an element of $\Theta(\mathcal{F})$ such that $\displaystyle\bigcup_{F \in Feat(SP)} Sig(F) \subseteq Sig(\overline{SP})$.

  We say that the integration of the features $Feat(SP)$ according to SP introduces feature interactions if there exists a feature F in $Feat(F)$ such that $\overline{SP} \models Spe(F)$ does not hold (i.e. SP does not preserve all the properties required by the features occurring in it).

Let us first remark that for us, specifications describe a set of models over a given signature: this set represents all the acceptable implementations of the specification. On the contrary, features are only defined by a set of expected properties. The underlying motivation of such a differentiated presentation is that a feature by itself is rarely understood as a complete description of a whole system, just as the (partial) description of an optional part of the system. The connection between specifications and features is that in practice, one imposes that some particular specifications resulting from the integration of a unique feature should preserve the properties required by the feature: it is the case for the first step of integration when a feature is plugged on some suitable basic specification (for example, specifying a basic telecommunication system). In section 4, we give some examples of such composition operators. In the sequel, for simplicity, when it is possible, we will sometimes consider terms of $\Theta(\mathcal{F})$ as the specifications they interpret.

Usually, in an algebraic setting, a basic specification SP is often simply defined as a simple axiomatic specification by means of a signature Σ and a set of Σ-sentences Ax, called axioms. More complex specifications are defined by combining some simpler specifications by means of some structuration primitives [20]. Each structuration primitive allows to fully characterise the resulting specification. For example, if we consider the classical sum primitive $+$ allowing to make the union of two existing specifications SP_1 and SP_2 defined on a common signature Σ, then the resulting specification $SP_1 + SP_2$ is defined by its signature Σ and by its set of models $\{M \mid M \in Mod(\Sigma), M \in Mod(SP_1), M \in Mod(SP_2)\}$. In case of feature-oriented systems, while there is a general agreement to say that such systems are defined by the way the integration of features has been dealt with, there is no agreement to highlight some precise feature combination operators (see section 1). Even more, whatever are these combination operators, the integration of features remains unpredictable or at least, not necessarily suitable with respect to the informal requirements. Indeed, as features define some precise new behaviours on some common underlying basic system, the modifications they all involve separately may introduce feature interactions when they are combined together. Indeed, since the properties required by different features may be conflicting, the integration of features cannot preserve all the involved features : the resulting integration expresses some choices between all the required properties by preserving some of them, but also by deleting or modifying the others. Typically, these choices are the concern of the specific knowledge of the application domain (here telecommunication systems):

thus, experts put in place a feature integration, having in mind the properties they want to preserve or to delete. Nevertheless, since the integration often means a very intricate combination of features (for example by writing ad hoc programming code), it remains for the expert to verify whether the integration of features as it has been designed meets the intended properties that the expert is thinking about. So, the design of a feature-oriented system amounts to have two tasks in hand: to make some choices for feature integration (for example, to put some priority mechanisms between features) and to make clear the expected properties of the resulting system. More or less, this last task consists in anticipating where are the feature interactions, and in classifying them as undesirable or desirable.

2.2   Feature design context

Following the discussion developed in the previous section, we model the activity of designing a feature-oriented system as the given of a pair whose :

- first element is the description of the integration of features under the form of a specification
- second element corresponds to the properties which are assigned to be significant for the system under design.

These two parts are given by experts but they are not necessarily directly connected. Indeed, even if these parts are often inter-defined in the sense that experts design each part taking into account the existence of the other part, nothing requires that the specification resulting from the integration of features satisfies the intended properties as they are classified. In our framework, the search for interactions would more or less consist in comparing the specification w.r.t the intended properties. This approach presents the advantage to be rather flexible for the experts in charge of the design of feature-oriented systems: they can separately adjust the integration of features, and the characterisation of the properties they expect for the resulting system. We now introduce the notion of feature design context capturing this notion of a coupling between feature integration and intended properties of the resulting system:

Definition 3
Let Σ be a signature of Sig. A feature design context C over Σ is a pair $< SP, IP >$ where

- *SP, also denoted $Sp(C)$, is a specification over Σ,*
- *IP, also denoted $\mathcal{I}(C)$ characterises the intended properties of the system under design as a triple $(\Lambda, \Delta, \Upsilon) \subseteq Sen(Sig)^3$ where:*
  - *Λ, also denoted $\mathcal{E}(C)$, is a set of expected properties,*
  - *Δ, also denoted $\mathcal{D}(C)$, is a set of declined properties,*
  - *Υ, also denoted $\mathcal{U}(C)$, is a set of undifferentiated properties.*

Let us point out that most of the times, the specification SP occurring in a feature design context C belongs to $\Theta(\mathcal{F})$. Intuitively, it simply means that SP results from the integration of the features $Feat(SP)$. With the notations of the definition 3, and according to the choices made by the expert to elaborate the considered feature design context :

- Λ may represent the desirable interactions (or suitable interworking interactions) of SP, the feature-oriented system under design ;
- Δ may represent the undesirable interactions (or rejected interactions) ;
- Υ represent all the properties that the expert cannot currently state whether they are desired or not. In fact, it means that the expert should investigate to know how the specification deals with them and/or to decide if he considers them as desirable or not. In other words, Υ represent all the properties still remaining to study in order to classify them. Thus, if C' denotes the feature design context built in the next step of the process, the expert can choose to incorporate them or only a part of them in $\mathcal{E}(C')$, or $\mathcal{D}(C')$ or else $\mathcal{U}(C')$. The expert can even choose to delete them from C' if finally, it does not matter to the expert how the specification behaves with respect to these properties. On the contrary, it could be interesting to reject properties by enforcing their negation: instead of placing a property in $\mathcal{D}(C)$, on can prefer to put its negation in $\mathcal{E}(C)$, especially when this property is considered as to be undesirable.

The definition 3 is very general and calls for some comments. The notion of feature design context is useful for describing a whole process of designing a system by integrating features. Each design step is characterised by a context fixing which are the current specification and the intended properties of the system. Then, from the careful study of the context and from their knowledge, the experts modify the context in order to tune both the feature integration and the intended properties with respect to the ideal system they have in mind. To start the process, a good starting point of what we call the "feature design process" may be a feature design context of the form $< SP, (\emptyset, \emptyset, \bigcup_{F \in Feat(SP)} Spe(F)) >$ with SP in $\Theta(\mathcal{F})$. To firstly choose how to compose the features of $Feat(SP)$, it is possible to fix some privileged composition operators in Θ and to systematically use them in the beginning of the process. Of course, an expert may decide to start from another specific context depending on his specific knowledge of the system under design. As a matter of fact, the goal is to refine an initial context in order to get an adequate and conclusive feature design context, each step being submitted to expert decisions. Let us define what these notions mean.

Definitions 4
Let C be a feature design context.

- *C is adequate means that the specification of C satisfies the expected properties of C, i.e. $Sp(C) \models \mathcal{E}(C)$ and does not impose the declined*

properties of C, i.e. $\forall \delta \in \Delta, \exists M \in Mod(Sp(C)), M \not\models \delta$ and that the set of SP-models is not empty, i.e. $Mod(SP) \neq \emptyset$.
- C is conclusive means that $\mathcal{U}(C) = \emptyset$.

3   Iterative process for designing feature-oriented systems

3.1   Design preorder on feature design contexts

In this section, we define a design step as a relation between feature design contexts:

Definition 5
Let C_1 and C_2 be two feature design contexts. We say that there is a design step from C_1 to C_2, and we note $C_1 \leq C_2$ if and only if :

- $Sig(Sp(C_1)) = Sig(Sp(C_2))$
- $\mathcal{E}(C_1) \subseteq \mathcal{E}(C_2)$, $\mathcal{D}(C_1) \subseteq \mathcal{D}(C_2)$, and $\mathcal{U}(C_2) \subseteq \mathcal{U}(C_1)$

Let us remark that intuitively, a design step consists in better delimiting the intended properties of the system under design. The idea is that the process for designing a feature-oriented system is based on an incremental adjustment of the required properties on the global system. Clearly, the relation $\leq$ over the feature design contexts is an preorder relation[2]. Thus, it allows to build feature design contexts incrementally by accumulating knowledge about the intended properties of the system under design : a design process never re-opens the design decisions previously done about the status of a property as expected or declined. It only increases the set of expected or declined properties while eliminating or classifying undifferentiated properties. Thus, a feature-oriented design process is formally defined as follows:

Definition 6
A feature-oriented design process is an ordered sequence of feature design contexts $(C_i)_{i=1,n}$.
   A feature-oriented design process is said to be complete if the last feature design context is both adequate and conclusive.

Let us remark that we do not impose any correlation between the two specifications involved in a given design step. It may be explained that each design step may question the previous feature integration by proposing a new way to integrate the features (possibly by changing the involved combination operators). It relies on subjective decisions. So, a general idea is to provide the expert with a set of composition operators as various as possible such that these operators may be used by the expert to influence his choice. Ideally, the

---

[2] It is not anti-symmetric.

composition operators should be defined by means of general rules likely to be applicable for a large class of features. Indeed, we suppose that the expert have some minimal knowledge about the way that the considered features have to be composed. Usually, the definitions of the first composition operators are guided from classical interaction schemes. Moreover any interesting feature construct already introduced in previous works should of course be incorporated as far as possible in such a family of composition operators.

Even if the notion of process allows the designer to continuously revise the specification of the feature-oriented system, a feature-oriented design process may nevertheless contain several consecutive contexts with a common shared specification. It would simply mean that properties are incrementally classified as expected or declined using a unique specification as reference to help the expert to decide on the status of the properties. Indeed, as soon as the specification is executable (via some animation techniques, or via the description of a precise model), then the expert can use the specification as a prototype of the feature-oriented system under consideration. By performing the specification on some pertinent properties, the observation of the resulting behaviours should help the expert to take a decision, and thus to make the feature design process move forward. From a practical point of view, this could be done using testing methods. This is developed in the next section.

Regarding to the methodological point of view, we have proposed to model the design of feature-oriented systems as an incremental process based on expert decisions and yielding a fine description of the system under the form of an association between a specification of the system and intended properties of the system. This is rather original with respect to usual design methods applied to classical software which are not modified by the integration of new features. So, in order to fully model a development life cycle for feature-oriented systems, our methodological proposition should be validated by case studies of significant size (see [2] for a small pedagogical toy example), combined with classical software engineering methods as refinement, proving, testing, . . . and to really be applicable, assisted by tools as much as possible.

3.2   Analogies with specification-based testing methods

Since from our point of view, the settling of a feature-oriented system is both incremental and assisted by an expert judgement, our approach is clearly a pragmatic one. Intuitively, this way of proceeding offers clear analogies with testing methods. Firstly, let us give some intuitions about testing according to the the three main testing activities which are selection, submission and verdict.

Specification-based testing consists in selecting test cases from the specification and according to some criteria. These criteria allow to select test sets which correspond to some coverage notion defined on the specification. The main principle underlying most testing methods is the following: since testing amounts to execute the system under verification with only a finite

data set, one should carefully choose them. That is exactly the rôle of testing criteria: more or less, they allow to decompose an infinite domain into a
family of subdomains. From such a decomposition, one reasonably considers
that any element of each subdomain is a representative element of the whole
subdomain with respect to the testing activity. In other words, by selecting
an element in each sub-domain in order to build a test data set, the tester
does as if it is equivalent to say that the execution of this selected element
behaves correctly with respect to the specification or to say that if they would
be performed by the program, all the executions of the elements of the subdomain behave correctly. Of course, this assertion is only a working assumption.
Testing criteria are judged all the more pertinent since in practice this assumption is often true. This is a pure subjective point of view. Nevertheless,
such a process allows the tester to stress the qualitative rôle of criteria in
order to find a good compromise between the test set size and the confidence
degree that most of errors have been discovered during the testing step.

Once a test set has been selected, one should submit each test case to
the program under test and collect all the computed outputs. This step often
requires a special instrumentation devoted to execute the program under test
with the adequate input data.

The last step concerns is the verification one: does the program results
match the results expected from the specification ? This last step is often
recognised as difficult[3]. A way to simplify it is to prevent from these difficulties from the beginning of the testing process: the only selected test cases are
the ones that one can easily decide as successful or not when they are performed on a program. From a theoretical point of view, it suffices to consider
that a test case contains the informations of both the selected input data t_i
and the expected output r for the function f under test: for example, it may
be modelled by an equation $f(t_1, \ldots, t_n) = r$. More generally, test cases are
simply properties (sentences) deduced from the specification and likely to be
interpreted as true or false by the program under test [1].

We now give some clues on how the feature design process as we have
defined it could be managed in practice using a pragmatic method similar
to the testing one. First, the description of the feature-oriented system by a
specification integration serves as the reference specification for testing. Second, the classification of properties (potential interactions) as desirable or
not is based on an expert judgement. With respect to a given property of the
undifferentiated property set, a natural way to help the expert to take his
decision is to assist him according to the following process: the first step consists in selecting some pertinent instances of the property according to some
criteria; the second one in using the reference specification to bring decision
elements for the expert. For example, if the instance is a simple scenario with

---

[3] A testing verdict may be established provided that computed results, expected
results, and decision procedures are available. These conditions are closely related
to observability notions.

the expected informations of the resulting state, then animating the specification may help the expert to decide whether he considers this instance as desirable or not. But to make the methodology really applicable, it should be asked to the expert to accompany his verdict with an explanation at each time he classifies a scenario as expected or declined. This explanation will be used to infer subdomains encompassing these scenarios: the underlying motivation is that a particular scenario, classified as expected or declined, is in fact representative of a larger class of scenarios having the same status with respect to the classification. Thus, the explanation may be useful to make a generalisation schema from the given selected scenario toward a whole subdomain. When comparing feature and testing domain, the strong connection between a particular instance and a whole subdomain relies respectively on the reliability of the expert judgement and on the quality of the selection criteria. Both of them are by nature subjective.

Finally, a typical design step in case of feature-oriented systems may be schematized as follows. Let us suppose that an undifferentiated sentence may be represented by a set of instances. Then the selection of a finite set of instances $\{s_i\}_{i \in I}$ is performed using coverage criteria similar to the ones used in testing methods. From the explanation given by the expert and associated to each selected instance, a set D_i of scenarios having the same status that the expert has decided for s_i may be considered. If these sets D_i may also be modelled by sentences, then at the next step of the design process, $D_{\mathcal{E}} = \cup_i D_i$ such that s_i has been accepted by the expert would increase the current set $\mathcal{E}(C)$ and similarly, $D_{\mathcal{D}} = \cup_j D_j$ such that s_j has been rejected by the expert would increase the current set $\mathcal{D}(C)$. According to the expert opinion, $D - D_{\mathcal{E}} - D_{\mathcal{D}}$ would be either remain in $\mathcal{U}(C)$ or eliminate from $\mathcal{I}(C)$ if it does not present an interest for the expert. Let us remark that the changing from a context C to a context C' may not involve the changing of the reference specification $Sp(C)$ as soon as the animation of the scenarios according to this specification remains useful to facilitate the classification of the properties by the expert. But of course, if this specification is far from meeting the expert requirements, then obviously, it would not so efficient to help the specifier to take his decision.

4   A simple framework to specify features

We now illustrate the different notions mentioned in the section 2. For this, we consider specifications written in a STR (state transition rule) style which is often used in the field of telecommunication systems (see [21], [17] or [10]). Basically, the behaviour of the system is described by means of rules expressing under which conditions an event may occur and which are its effects on the system state.

4.1   An example of a logic

We now give successively the different elements composing the considered logic, as defined in the section 2. This logic is very similar to the rule-based one used in [17]. Generally speaking, we keep the same notations. We just add the possibility to express some static properties on the states as simple propositional sentences.

Definition 7

Sig is the set of signatures where each signature Σ is defined as a 4-uple (U, V, P, E) where U is a set of constants, V is a set of variables, P and E are respectively a set of predicate symbols and a set of event symbols, all these symbols being provided with a natural number as arity.

Two signatures $\Sigma_1 = (U_1, V_1, P_1, E_1)$ and $\Sigma_2 = (U_2, V_2, P_2, E_2)$ verify $\Sigma_1 \subseteq \Sigma_2$ if and only if $U_1 \subseteq U_2$, $V_1 \subseteq V_2$, $P_1 \subseteq P_2$ and $E_1 \subseteq E_2$ where here, $\subseteq$ denotes the usual set inclusion.

Intuitively, U represents the set of users, E denotes the different events which make the system evolve and P allows the specifier to describe the different states of the system. Given a signature $\Sigma = (U, V, P, E)$, P_Σ (resp. E_Σ) denotes the set of all well-formed terms over P (resp. E), i.e. of all terms $p(t_1, \ldots, t_n)$ (resp. $e(t_1, \ldots, t_n)$) where p (resp. e) is a predicate of P (resp. event of E) of arity n and $t_1, \ldots, t_n$ are either elements of U or elements of V. We will denote C_Σ the set of all conditions over the signature Σ consisting in a set of predicate terms over Σ or of negations of predicate terms. For each condition c in C_Σ, $pos(c)$ will denote the subset of all positive elements of c while $neg(\Sigma)$ will denote the subset of all negative elements of c.

The sentences are either rules labelled by an event term or static properties carrying over states. These last sentences may be particularly useful to express safety properties which should be verified by the system under specification.

Definition 8

Given a signature $\Sigma = (U, V, P, E)$, the sentences of $Sen(\Sigma)$ are:

- *either rules of the form r : $pre[ev]post$ where r is the rule name, pre and post are conditions over Σ (in the sequel, given a rule r, the conditions pre and post will be denoted by $pre[r]$ and $post[r]$ and the event ev by $ev[r]$);*
- *or static properties which are given as usual propositional sentences over predicate terms and usual logical connectors.*

In order to define specifications, models and the satisfaction relation, we begin to introduce some useful preliminary notions defined as usual :

Definition 9

Let $\Sigma = (U, V, P, E)$ be a signature.

- A substitution $\theta : V \to U$ is a partial function with a finite set as domain. As usual, any substitution θ may be extended to the set of (predicate or event) terms overs V.
- A state s over Σ is defined as a set of ground predicate terms (i.e. terms without variables). The set of states over Σ is denoted by S_Σ.
- A state s satisfies a predicate term $p(t_1, \ldots, t_n)$ if and only if for each substitution θ, s contains $p(\theta(t_1), \ldots, \theta(t_n))$. More generally, s satisfies a condition c if and only if for any positive (resp. negative) element v of c, s satisfies v (resp. does not satisfy v). The satisfaction of a static property by a state is defined as usual for propositional sentences according to the truth tables of the logical connectors. More generally, a condition c satisfies a condition c' if for every substitution θ, for every state s satisfying $\theta(c)$, s also satisfies $\theta(c')$.

Definition 10
Let $\Sigma = (U, V, P, E)$ be a signature. A specification $SP = (Ax, s_0)$ over Σ is given by a set Ax of sentences over Σ and an initial state s_0 in S_Σ.

Definition 11
Let $\Sigma = (U, V, P, E)$ be a signature. The set of Σ-models $Mod(\Sigma)$ is the set of all automatas $M = (S, T, s_M)$ where:

- S is a subset of S_Σ;
- T is a subset of $S_\Sigma \times E_\Sigma \times S_\Sigma$ such that for any state s and event term e, there exists at most an (s, e, s') in T. Each element (s, e, s') of T is called a transition;
- s_M is the initial state of M and is an element of S_Σ.

Definition 12
Let $\Sigma = (U, V, P, E)$ be a signature, $M = (S, T, s_M)$ a Σ-model and ϕ a Σ-sentence. The model M satisfies the sentence ϕ if and only if:

- if ϕ is a rule of the form $pre[ev]post$, then for every substitution θ, for every state s of M satisfying $\theta(pre)$, there exists a transition $(s, \theta(ev), s')$ in T verifying[4] $s' = s - \theta(pre) + \theta(pos(post)) - \theta(neg(post))$.
- if ϕ is a static property, then for every state s of M, s satisfies ϕ as a propositional sentence.

Let $SP = (Ax, s_0)$ a specification. M satisfies SP if and only if M satisfies each axiom ax of Ax and s_M contains the initial state s_0 of SP.

---

[4] As in [17], $+$ and $-$ are respectively the subtraction and union operations on the sets.

4.2   The set of composition operators

We give now three classes of composition operators. The first one represents the basic case introducing the first properties to be considered. More or less, it amounts to recopy the axioms of the feature given in argument. In particular, such an operator allows the specifier to introduce the description of the underlying base system under the form of a feature. The second one consists in restricting the scope of the already existing axioms verifying some given condition. It allows to prevent some foreseeable conflicts between properties. The last class allows to describe some priority mechanisms between features: some properties of a given feature can be deleted provided that they match some conditions. Intuitively, there should exist some corresponding properties which are considered as having a higher priority level with respect to the properties to be deleted. The two last operators are parameterised by some parameters as conditions, events, ... Thus, it allows the specifier to precisely define the integration of features according to his needs.

Definition 13
The set Θ of composition operators is defined as the union of the three sets Θ_B, Θ_R and Θ_D where:

- *Θ_B contains one operator of arity 1 denoted by the identifier b ;*
- *Θ_R is the set of operators of arity 1 which are generically denoted as $rest_{c,e,r}$ where c is a condition, e an event term and r a condition, ($\Sigma_{s,e,r}$ denotes the union of all signatures involving for writing c, e and r);*
- *Θ_D is the set of operators of arity 2 which are generically denoted as $del_{<s_j,e_j>_{i \in J}}$ where c_j is a condition, e_j an event and J a finite index ($\Sigma_{<s_j,e_j>_{i \in J}}$ denotes the union of all signatures involving for writing s_j, e_j with j in J).*

For simplicity, if the index J is empty (resp. reduced to a singleton), the operator will be simply denoted del (resp. $del_{c,e}$). Given $\mathcal{F}$ a set of features as defined in section 2, we now give the interpretation of the terms of $\Theta(\mathcal{F})$ as axiomatic specifications:

Definition 14
The interpretation of a term t of $\Theta(\mathcal{F})$ is inductively defined according to its form:

- *if t is written F with F a feature in $\mathcal{F}$ with Σ and with its set $Spe(F)$ of specific properties, then $\overline{F}$ is interpreted as the axiomatic specification whose signature is Σ and set of axioms is $Spe(F)$;*
- *if t is written $b(t')$ with t' a term of $\Theta(\mathcal{F})$ whose interpretation gives Σ' as signature and Ax' as set of axioms, then $b(t')$ is interpreted as the the axiomatic specification as for t';*

- if t is written $rest_{c,e,r}(t')$ with t' in $\Theta(\mathcal{F})$ with Σ' as signature and Ax' as set of axioms when t' is interpreted, then the interpretation of t has $\Sigma' \cup \Sigma_{c,e,r}$ as signature and $(Ax' - \{\phi \mid \phi \in Ax', c \models pre[\phi], ev[\phi] = e\}) \cup \{pre[\phi] + r[e]post[\phi] \mid \phi \in Ax', c \models pre[\phi], ev[\phi] = e\}$ as set of axioms.

- if t is written $del_{<c_j,e_j>_{j\in J}}(t_1, t_2)$ with for i=1,2 t_i in $\Theta(\mathcal{F})$ with for respective interpretations, Σ_i as signature and Ax_i as set of axioms, then t has $\Sigma_1 \cup \Sigma_2 \cup \Sigma_{<c_j,e_j>_{j\in J}}$ as signature and $(Ax_1 \cup Ax_2 - \{\phi_1^j \mid \phi_1^j \in Ax_1, c_j \models pre[\phi_1^j], ev[\phi_1^j] = e_j, \exists \phi_2^j \in Ax_2, c_j \models pre[\phi_2^j], ev[\phi_2^j] = e_j\}$ as set of axioms.

Let us remark that the interpretation of *del* (without any index) obviously corresponds to the sum (or union) of two specifications as mentioned in section 2. These composition operators may seem rather ad hoc, but they precisely allow the specifier to adjust the definition of the integration of features according to some precise requirements. Let us point out that in order to fit in with more generality, the definition of such operators should be defined up to variable renaming: for simplicity, we make the assumption that the same variables are always chosen for the same rôle: in practice, such an assumption is not realistic but it presents the advantage of simplifying the presentation of the example. This set of operators only represents a first base in order to be able to specify a feature-oriented system. Of course, one can consider any other systematic or ad hoc composition operators if they correspond to some needs.

5   An example of an iterative process

We illustrate our purpose using the logic given in the previous section on the classical example already discussed in the introduction and concerning the well-known features OCS and CFU. The basic system ($POTS$) and the two features are uniformly given under the form of features.

5.1   Description of the features

The basic system $POTS$ is described as a feature whose meaning of properties is rather clear from the choice of identifiers:

- Σ_{POTS} = { U = { tel_1, tel_2, tel_3, tel_4 }, V = {x, y, z}, P = { (idle, 1), (linebusy, 1), (log, 2) }, E = { (Dial, 2), (Onhook, 1) }}
- $Spe(POTS)$ consists of the following sentences :
  ϕ_1^{POTS} :{idle(x), idle(y)} [Dial(x, y)] {log(x, y)), log(y, x)},
  ϕ_2^{POTS} : {idle(x), ¬ idle(y)} [Dial(x, y)] {linebusy(x)},
  ϕ_3^{POTS} : {log(x, y), log(y, x)} [Onhook(x)]{idle(x)), idle(y))}),
  ϕ_4^{POTS} : {linebusy(x)} [Onhook(x)] {idle(x))},
  ϕ_5^{POTS} : linebusy(x) ∨ log(x,y) ⇔ ¬ idle(x),

Of course, the CFU feature is also described as a feature:

- $\Sigma_{CFU} = \{U = \{ tel_1, tel_2, tel_3, tel_4 \}, V = \{x, y, z\}, P = \{ (idle, 1),$
  (linebusy, 1), (log, 2), (CFU, 2) \}, E = \{ (Dial, 2), (Onhook, 1) \}\}$
- $Spe(CFU)$ consists of the following sentences :
  ϕ_1^{CFU} : \{CFU(y, z), idle(x), idle(z)\} [Dial(x, y)]
  \hspace{3em} \{CFU(y, z), log(x, z), log(z, x)\},
  ϕ_2^{CFU} : \{CFU(y, z), idle(x), ¬ idle(z)\} [Dial(x, y)]
  \hspace{3em} \{CFU(y, z), linebusy(x)\},

Lastly, the OCS feature is now given. The OCS(y,x) term indicates that the phone x is on the list of the phone y subscribing the OCS feature.

- $\Sigma_{OCS} = \{ U = \{ tel_1, tel_2, tel_3, tel_4 \}, V = \{x, y, z\}, P = \{(idle, 1),$
  (linebusy, 1), (log, 2), (OCS, 2)\}, E = \{(Dial, 2), (Onhook, 1)\}\}$
- $Spe(OCS)$ consists of the following sentence :
  ϕ_1^{OCS}: \{OCS(y, x), idle(x), idle(y)\} [Dial(x, y)]
  \hspace{3em} \{OCS(y, x), linebusy(x)\},
  ϕ_2^{OCS}: (OCS(y, x) ∧ ¬ log(x, y)) ∨ ¬ OCS(y, x),

5.2   An example of incremental integration of features

In this section, we describe an example of the step by step process of feature integration for the three features (POTS,CFU,OCS) given above. We will stress on the active rôle of the expert in the setting of the resulting specification of the feature-oriented system.

Step 1: the starting point At the beginning of the integration process, the expert may decide to use the union mechanism as composition operators. Of course, the expert suspects that it will not be the right final feature integration. Thus, if we note SP_1 the term $del(b(CFU), del(b(POTS), b(OCS)))$ of $\Theta(\mathcal{F})$, we choose as the initial context the following one:

$$\mathcal{C}_1 = \, < SP_1, (\emptyset, \emptyset, \bigcup_{F \in Feat(SP_1)} Spe(F)) >.$$

Step 2: a classification step Among all the properties required by the involved features, the expert may wish to preferably preserve some of them. It is a pure subjective part guided by the knowledge of the features under consideration. Such a choice cannot be done systematically and may be viewed as a design choice. Here, besides the fact that the expert would keep the static property given in the POTS description, the expert may decide that the second OCS property should be preserved during the integration process. So, the expert gives a second feature design context where ϕ_5^{POTS} and ϕ_2^{OCS} are both added to the set of expected properties and deleted from the set of undifferentiated properties. So, the new design step consists in:

$$\mathcal{C}_2 = \, < SP_1, (\{\phi_5^{POTS}, \phi_2^{OCS}\}, \emptyset, \bigcup_{F \in Feat(SP_1)} Spe(F) - \{\phi_2^{OCS}, \phi_5^{POTS}\}) >.$$

Clearly, we have $\mathcal{C}_1 \leq \mathcal{C}_2$.

Step 3: a new composition If we put the OCS property ϕ_2^{OCS} in disjunctive normal form, then we get three subcases which are respectively { OCS(y,x),$\neg$ log(x,y)}, {$\neg$ OCS(y,x),log(x,y)} and {$\neg$ OCS(y,x),$\neg$ log(x,y)}. As this property ϕ_2^{OCS} is clearly conflicting with properties of POTS, in particular with the property ϕ_1^{POTS}, the idea is to combine the three different subcases issued from the decomposition of ϕ_2^{OCS} with the pre-condition of ϕ_1^{POTS} in order to make appear some relevant subcases of ϕ_1^{POTS} that the expert could decide about. Here, it will normally give the following subcases: { OCS(y,x),$\neg$ log(x,y),idle(x),idle(y)},   {$\neg$ OCS(y,x),log(x,y),idle(x),idle(y)} and {$\neg$ OCS(y,x),$\neg$ log(x,y),idle(x),idle(y)}. The first case is in contradiction with ϕ_5^{POTS} while the two others may be simplified by suppressing $\neg$ log(x,y). So, in the context of the property ϕ_2^{OCS}, the rule ϕ_1^{POTS} may be decomposed in two rules, adding in its precondition predicates terms corresponding to the two pertinent subcases. Let us remark that these added predicates correspond to subscription predicates. They are also added to the postcondition part of the rule in order to preserve the rule satisfaction definition. Indeed, the case analysis should not modify the satisfaction mechanisms. Finally, we obtain the two following rules:

ϕ_{11}^{POTS}: {idle(x), idle(y), OCS(x, y)} [Dial(x, y)]

$\qquad$ {log(x, y), log(y, x), OCS(x,y)}

ϕ_{12}^{POTS}: {idle(x), idle(y), $\neg$ OCS(x, y)} [Dial(x, y)]

$\qquad$ {linebusy(x), log(x, y), log(y, x), $\neg$ OCS(x,y)}

It amounts to make a case analysis with respect to the required static properties (here, ϕ_2^{OCS}) and to carry it to other rules. It allows to subdivide properties in order to discover subcases for which the expert is likely to decide in a systematic way. Here, by keeping the rule ϕ_{11}^{POTS}, it would clearly lead to a contradiction with the property ϕ_2^{OCS}. So, a rather natural decision consists in restricting the application of ϕ_1^{POTS} to the case where there is no OCS subscription. The same situation occurs with respect to the CFU feature. So, the expert could choose to give a more subtle feature integration where properties are restricted to the cases without subscriptions. For this, we consider the condition c_1 as {idle(x), idle(y), OCS(y, x)} and the condition c_2 as {idle(x),idle(y), idle(z), CFU(y,z)} which satisfy both preconditions of POTS rule according to the event Dial(x,y) and of OCS or CFU rules. With these notations, the expert gives a third feature design context where the reference specification has changed. Let SP_2 be the specification term where e_1 is the event term Dial(x,y):

$del(rest_{\neg CFU(y,z),c_2,e_1}(del(rest_{\neg OCS(y,x),c_1,e_1}(b(POTS)),b(OCS))),$

$\qquad b(CFU))$.

The expert can then put in place the following context:

$\mathcal{C}_3 = \; < SP_2,(\{\phi_5^{POTS},\phi_2^{OCS}\},\emptyset, \; \bigcup_{F\in Feat(SP_1)} \; Spe(F) - \{\phi_2^{OCS},\phi_5^{POTS}\}) >.$

Step 4: a new classification step : the expert now decides to make a new case analysis w.r.t the subscription terms OCS(y,x) and CFU(y,z) and to superimpose it on the property ϕ_1^{POTS}. It gives the 4 following subcases:

ϕ_{111}^{POTS}: {idle(x), idle(y), OCS(y, x), CFU(y,z)} [Dial(x, y)]

{log(x, y), log(y, x), OCS(y, x), CFU(y,z)}

ϕ_{121}^{POTS}: {idle(x), idle(y), ¬ OCS(y, x), CFU(y,z)} [Dial(x, y)]

{log(x, y), log(y, x), ¬ OCS(y, x), CFU(y,z)}

ϕ_{112}^{POTS}: {idle(x), idle(y), OCS(y, x), ¬ CFU(y,z)} [Dial(x, y)]

{log(x, y), log(y, x), OCS(y, x), ¬ CFU(y,z)}

ϕ_{122}^{POTS}: {idle(x), idle(y), ¬ OCS(y, x), ¬ CFU(y,z)} [Dial(x, y)]

{log(x, y), log(y, x), ¬ OCS(y, x), ¬ CFU(y,z)}

The expert decides to classify them as expected or declined: he declines ϕ_{111}^{POTS} and ϕ_{112}^{POTS} grouped together in ϕ_{11}^{POTS} and he keeps ϕ_{122}^{POTS} as expected. For ϕ_{121}^{CFU}, he remarks that the precondition matches with a precondition of a CFU rule sharing the same event. So, he also prefers to decline it.

$$C_4 = \; < SP_2, (\{\phi_5^{POTS}, \phi_2^{OCS}, \phi_{122}^{POTS}\}, \{\phi_{11}^{POTS}, \phi_{121}^{CFU}\},$$
$$\bigcup_{F \in Feat(SP_2)} Spe(F) - \{\phi_2^{OCS}, \phi_5^{POTS}, \phi_1^{POTS}\}) >.$$

Step 5: a composition step : the expert chooses to decompose the ϕ_1^{CFU} rule with respect to the fact that either OCS(z,x) holds or it does not.

ϕ_{11}^{CFU}: {CFU(y, z), idle(x), idle(z), OCS(z, x)} [Dial(x, y)]

{CFU(y, z), log(x, z), log(z, x)}

ϕ_{12}^{CFU} = {CFU(y, z), idle(x), idle(z), ¬ OCS(z, x)} [Dial(x, y)]

{CFU(y, z), log(x, z), log(z, x)}

We can observe that the rule ϕ_{11}^{CFU} would lead to the non-satisfaction of the property ϕ_1^{OCS} property which is expected. This is exactly corresponds to the interaction case described in the introduction. So, the expert decides to propose a new feature composition consisting in restricting the scope of the ϕ_{11}^{CFU} rule. If we note c_3 the condition {CFU(y, z), idle(x), idle(z), OCS(z, x)}, we can consider the specification with e_1 denoting the Dial(x,y) event:

$$SP_3 = del(del(rest_{\neg CFU(y,z),c_3,e1}(rest_{\neg OCS(y,x),c_1,e1}(b(POTS))), b(OCS)),$$
$$rest_{\neg OCS(z,x),c_3,e1}(b(CFU)))$$
$$C_5 = \; < SP_3, (\{\phi_5^{POTS}, \phi_2^{OCS}, \phi_{122}^{POTS}\}, \{\phi_{11}^{POTS}, \phi_{121}^{CFU}\},$$
$$\bigcup_{F \in Feat(SP_2)} Spe(F) - \{\phi_2^{OCS}, \phi_5^{POTS}, \phi_1^{POTS}\}) >.$$

Step n: ... the process may continue until the expert is fully satisfied with the final feature design context which ideally should be both adequate and conclusive.

Of course, our method should be assisted by tools in order to be fully efficient. The two major steps to be assisted by tools are the one of the verification of a given property by a specification and the one of the decomposition of a given property into a family of properties covering the whole scope of the initial property, this decomposition being done according to some criterias as the satisfaction of a static property. VALISERV is a new French project concerning features in telecommunication systems and including four organisations : France Telecom R&D, LaMI (University of Evry), LSR-IMAG (University Joseph Fourier of Grenoble) and TNI, an industrial partner. Regarding to the feature interaction problem, the VALISERV approach is a pragmatic one and addresses the incremental design of feature-oriented systems using testing methods. The difficult point before designing appropriate tools is the choice of a formalism well-adapted for both feature specifications and for testing-like verification methods.

6   Conclusion

We have proposed a design method devoted to feature-oriented systems. When new features are integrated on a system already including features, it is widely recognised that the resulting system is unpredictable. So, we prone to design feature-oriented systems according to an incremental and pragmatic process. Each stage is modelled by a feature design context characterized by a specification resulting from the integration of some features and by a set of properties which are respectively classified as expected (interworking interactions), declined (bad interactions), or undifferentiated (properties still to be studied). Each design step is based on expert knowledge. At each step, the expert expresses some judgement on the qualitative nature of the undifferentiated properties, and possibly, adapt the definition of the reference specification in accordance to his decisions. The process ends when the set of undifferentiated properties becomes empty and when the specification verifies the intended properties of the final context.

Obviously, such a design method should be assisted by tools to be fully efficient. For this, we plan to be inspired by the techniques of specification-based testing. Indeed, testing and feature design share a major common point. For both methods, a subjective element plays a significant part. It is respectively the quality of testing criteria and the classification of interactions as desirable or not.

References

1. A. Arnould, P. Le Gall, and B. Marre. Dynamic testing from bounded data type specifications. In *Proc. of EDCC-2 Second European Dependdable Computing Conference, Taormina, Italy*, 1996.
2. G. Bernot, H. Jouve, F Klay, F Ouabdesselam, and J.-L. Richier. Aide à l'intégration de services par la génération de tests. In *AFADL'00, Grenoble, France*, 2000.
3. L.G. Bouma and H. Velthuijsen, editors. *Feature Interactions in Telecommunications and Software Systems (FIW'95)*. IOS Press, 1995.
4. L.G. Bouma and H. Velthuijsen, editors. *Feature Interactions in Telecommunications and Software Systems (FIW'98)*. IOS Press, 1998.

5. Jan Bredereke. Families of Formal Requirements in Telephone Switching. In *[6]*, pages 257–273, 2000.
6. M. Calder and E. Magill, editors. *Feature Interactions in Telecommunications and Software Systems (FIW'00)*. IOS Press, 2000.
7. P. Combes and S. Pickin. Formalization of a user view of network and services for feature interaction detection. In *Feature Interactions in Telecommunications Systems*, pages 120–135. IOS Press, 1994.
8. R. Diaconescu, J. Goguen, and P. Stefaneas. Logical support for modularization. In G. Huet and G. Plotkin, editors, *Logical Environments*, Proceedings of a Workshop on Logical Frameworks, pages 83–130, may 1991.
9. L. du Bousquet, F. Ouabdesselam, J.-L. Richier, and N. Zuanon. Feature interaction detection using synchronous approach and testing. *Computer Networks and ISDN Systems*, 1999.
10. A. Gammelgaard and J. E. Kristensen. Interaction detection, a logical approach. In L.G, Bouma and H. Velthuijsen, editors, *Feature Interactions in Telecommunications Systems*, pages 178–196. IOS Press, 1994.
11. C. Gaston, M. Aiguier, and P. Le Gall. Algebraic treatment of feature-oriented systems. Technical Report 44, LaMI, Université d'Evry, 2000.
12. J.P. Gibson. Towards a feature interaction algebra. In *[4]*, pages 217–231, 1998.
13. S. Gilmore and J. Hillston. Implementing the PEPA feature construct. In *Workshop on Language Constructs for Describing Features*, pages 23–38. Stephen Gilmore and Mark Ryan (editors),Glasgow, 2000.
14. J. A. Goguen and R. M. Burstall. Institutions: abstract model theory for specifications and programming. *association for Computing Machinery*, 1992.
15. M. Goldberg, E. Magill, D. Marples, and S. Reiff. Second feature interaction contest. In *[6]*, pages 293–310, 2000.
16. F. Klay, M. Rusinowitch, and S. Stratulat. Analysing Feature Interactions with Automated Deduction Systems. In *7th International Conference on Telecommunication Systems Modeling and Analysis*, 1999.
17. M. Nakamura, Y. Kakuda, and T. Kikuno. Feature Interaction Detection Using Permutation Symmetry. In *Feature Interactions in Telecommunications Systems V*, pages 187–201. IOS Press, 1998.
18. G. Nancy, B. Blumenthal, J.C. Gregoire, and T. Otha. Feature interaction detection contest. In *[4]*, pages 327–359, 1998.
19. M. Plath and M. Ryan. Plug-and-play features. In *[4]*, pages 150–164, 1998.
20. M. Wirsing. Structured Specifications: Syntax, Semantics and Proof Calculus. In Brauer W. Bauer F. and Schwichtenberg H., editors, *Logic and Algebra of Specification*, pages 411–442. Springer, 1993.
21. Yoneda-Ohta. A formal approach for definition and detection of feature interaction. In *[4]*, pages 202–216, 1998.
22. P. Zave. Architectural solutions to feature-interaction problems in telecommunications. In *[4]*, pages 10–22, 1998.

Abstraction and refinement of features

Dominique Cansell[1,2] and Dominique Méry[1,3]

[1] LORIA UMR 7503, BP 239, Campus Scientifique,54506 Vandœuvre-lès-Nancy
   Cédex, France
[2] Université de Metz, Ile du Saulcy, 57045 Metz Cédex, France
[3] Université Henri Poincaré Nancy 1, BP 239, Campus Scientifique, 54506
   Vandœuvre-lès-Nancy Cédex, France
   email: {cansell,mery}@loria.fr

Abstract. The composition of services and features often leads to unwanted situations, because it is a non-monotonic operation over services and features. When a new service is added to an existing system, conditions have to be checked to ensure that the resulting system satisfies a list of required properties. Following the *system* approach of Abrial, we develop services and features in an incremental way and use refinement to model the composition of services and features. Proof obligations state the preservation or the non-preservation of properties, namely invariant or more generally safety properties. The method helps us in understanding when a service is interfering with another, and allows us to give multiple views of each service according to the level of its refinement. Finally, we validate our method with the Atelier B tool.

1 Introduction

The composition of services and features is a non-monotonic operation, since it may not preserve properties of the involved services or features. The feature interaction problem in telecommunications and software engineering reports observations of undesirable behaviours with respect to service requirements, when composing services. Services requirements state properties such as safety, liveness or fairness; but the main requirements of telecommunications services fall into the safety category. We restrict our requirements to safety properties and we develop a methodology for analysing causes of feature interactions among services and features.

A service is defined as a finite sequence of views related by a refinement relationship, which allows one to improve the precision of the model. When one adds a new variable or when one adds a new operation, a property held by a previous view can be corrupted in the new view; the detection of the corruption of a previous property is permitted by the checking of proof obligations generated for every view. In fact, when a new view refines a previous one, proof obligations for expressing the refinement relationship between the two views are generated from the text of the two views: it is an internal validation of views. When proved, proof obligations ensure that the current view satisfies an invariant and other safety properties called assertions. A view of a service is a (formal) model of a service including (flexible) variables, an invariant property typing variables, an initial condition for variables and a finite set of events modelling the change of variables but preserving the invariant. Hence, a service is characterised by at least one view and other views are related by a refinement relationship.

The process of composition of services through the composition of views requires us to define how we compose views. We will use the services of CS 1 [32] for validating our approach and we first consider how those services extend the functionalities of basic call service called $BASE$. Following our modelling of services through *views*, $BASE$ is a sequence of views $(BASE_i)_{i \in I}$ and the composition of another service is defined by incrementing every view of $BASE$ by elements leading to views of the new service. The instantiation of a new service is traditionally represented using the $\oplus$ operator (see Combes et all [13]); $\oplus$ is a symbolic notation and we get a meaning for it through refinement. $BASE \oplus S$ is then defined by its views $(BASE_i \oplus S_i)_{i \in I}$ satisfying $\forall i \in I : BASE_i \sqsubseteq BASE_i \oplus S_i$: $\sqsubseteq$ expresses the refinement relationship over views. The refinement process is controlled by proof obligations and guarantees the preservation of safety properties of the currently developed service; it preserves abstract traces (or behaviours). The idea is clearly related to our previous work [25–27] on using B for modelling services and on the work of J.R. Abrial [2,3] on the B event-based approach - called B system. The incremental development of complex systems is made easier because of refinement, which appears to be a structuring mechanism for services. The $\oplus$ operator is given a meaning based on the refinement relationship. We have incrementally refined different views of systems. Our refinement is carried out in two main directions:

- making abstract models more concrete: we add more and more details that were considered as hidden in the more abstract models and we call this *the horizontal refinement.*
- instantiating a new service on the current system: we add new variables and new events for the new service and we call this *the vertical refinement.*

In [9], an incremental method for developing services allows us to build several views (or models) of the same system; views are related by a (formal) relationship called refinement and partly supported by a tool [30]. The refinement adds details of the informal description of the system and makes formal models more complex; adding a new service is not difficult but it may violate properties of system components. The refinement called vertical refinement is used to define the composition of a current system and a new service. The composition of a system and a new service is carried out by refining every view of the system. The resulting chain of refined models has to satisfy the constraint of horizontal refinement. Typically, we have a chain of views such as:

$$V_1 \xrightarrow{j_1} \ldots \xrightarrow{j_{n-1}} V_n$$

where $\xrightarrow{j_i}$ states the refinement of views and V_i is the view of a system at different stages of details.

The label over edges states the number of unprovable proof obligations; it may appear that a view is not the full refinement of another view, since it means that we detect an interaction. Now, we add a new service SERVICE into V and we get the diagram of figure 1 for every i in $\{1, \ldots, n-1\}$.

Fig. 1. Analysis Pattern for Refining Services

Our initial analysis suggested that the absence of interactions is defined by *npov* and *npoh* being equal to 0; we achieve a new system, which preserves properties of abstract models in both directions: preserved properties are safety properties. Another derived idea is to use both refinements for analysing behaviours of systems extended by services. We do not want to derive the implementation of the full phone system, because of many proof obligations, but we can use it for analysing requirements and preventing possible interactions, when developing services. Unprovable proof obligations help us to transform $(\rightsquigarrow)$ $V_{i+1} \oplus \text{SERVICE}_i$ into $V_{i+1} \oplus \text{SERVICE}_{i+1}$, which is the view we are planning to refine. We call this method the system-mate approach. We give a first introduction of the method in [9], but we have not detailed the process for obtaining a new abstract (model or) view through refinement, especially the use of unprovable proof obligations.

Classical approaches [21,24] to feature interaction problem are mainly based on an analysis of formal models (state-based, temporal, algebraic, semi-formal, relational ...) and the validation of formal models is carried out after writing specifications or programs. Feature integration [28] uses model checking for validation of features properties; incremental integration validation [15] uses the synchronous model Lustre and a testing tool for validation of feature properties; modular description of features and static detection of feature interaction [31]are mainly based on an architectural approach founded on LOTOS; the state-based approach for defining and detecting feature interactions [25,33,17], the relational approach for defining and detecting feature interactions [16] and the logical approach for defining and detecting feature interactions [6,18,5] are closely related; finally, the rchitecture-oriented approach [20] intends to develop a framework for composing feature with respect to a given architecture but it is not clearly based on mathematical model.

The paper is structured as follows. Section 2 introduces abstract models for services using B and contains a modelling of the basic service; it shows how the basic phone model is incrementally refined using the system-mate approach. Section 3 describes the method for integrating services using the system-mate approach and the analysis pattern of figure 1. Section 4 discusses technical questions related to the use of the tool and to the complexity of designed abstract models and concludes the document.

2   Modelling views of services

An abstract model for a service is defined by a list of variables - characterising the state of the service, an invariant (generally called an inductive invariant) - expressing properties satisfied by variables, and a list of events which are handled by the service. The central idea is to avoid writing complex abstract models and to use refinement for adding details or features to the abstract model under development. The refinement allows one to control the validity of the current abstract model and to trace decisions in the development. An abstract model for a service is also called an event system. We recall that such an event system models actions done by a user or triggered by a user or done by the system. We illustrate our ideas on the basic call service called $BASE$ plus services such as CF (call forwarding), OCS (originating call screening), TCS (terminating call screening). We do not review here the B method and B definitions [1], but we use Atelier B[30] to validate every derived abstract model. Note that required proof obligations are partially generated by Atelier B. Let us recall that an abstract model for a service is a view with more or less details. An effective service is a commercial product and its description is very simple. Our main contribution is the use of invariants for stating safety requirements; the combination of two (or more) services is achieved by transformations over abstract models. The two main features of a telephone system are the calling feature and the receipt feature. Moreover, our model must be extensible and flexible. Hence, a telephone system provides a set of functionalities (calling, receiving, ringing, billing, ...) and there are user-managed functionalities and system-managed functionalities. We promote an interactive style for building abstract models of services and we assume that there are two actors:

- the customer who is waiting for a formal model and who is an expert in services.
- the system-mate who is the writer of abstract models and who has to communicate with the customer.

The contract between the customer and the system-mate is produced by an answers/replies party, as we explained in our first work on the system-mate approach or the B system approach for services [9]; the system-mate asks his

customer's questions to get more hints on the system, he is currently developing. He must be sure that his abstract models fit the informal requirements
of the customer. The dialogue for requirements is an effective part of the
development of abstract models; it helps the specifier (system-mate) and the
customer to get an agreement. However, the customer may not to be able
to understand the language of the specifier and the specifier has to use the
language of the customer, when reporting his work.

Name	Syntax	Definition
Binary relation	$s \leftrightarrow t$	$\mathcal{P}(s \times t)$
Domain	$DOM(r)$	$\{a \mid a \in s \wedge \exists b.(b \in t \wedge a \mapsto b \in r)\}$
Codomain	$RAN(r)$	$DOM(r^{-1})$
Identity	$\mathbf{id}(s)$	$\{x \mapsto x \mid x \in s\}$
Restriction	$s \lhd r$	$\mathbf{id}(s); r$
Co-restriction	$r \rhd t$	$r; \mathbf{id}(t)$
Anti-restriction	$s \ntriangleleft r$	$(dom(r)-s) \lhd r$
Anti-co-restriction	$r \ntriangleright t$	$r \rhd (ran(r)-t)$
Partial function	$s \nrightarrow t$	$\{r \mid r \in s \leftrightarrow t \wedge (r^{-1}; r) \subseteq \mathbf{id}(t)\}$
Partial into function	$s \rightarrowtail t$	$\{f \mid f \in s \nrightarrow t \wedge f^{-1} \in t \nrightarrow s\}$

Fig. 2. Notations for abstract systems

A call is a pair between two persons and it is a very abstract view of the
states of the persons and what is relating persons. *PERSONS* is a non-empty
set of persons and it is the set of potential customers. Now, the two identified
events are defined as follows:

Creating_a_call =
  any $p1$, $p2$ **where**
  $p1 \in PERSONS \wedge p2 \in PERSONS \wedge p1 \mapsto p2 \notin CALLS$
  then $CALLS := CALLS \cup \{ p1 \mapsto p2 \}$ **end**;

Halting_a_call =
  any $p1$, $p2$ **where**
  $p1 \in PERSONS \wedge p2 \in PERSONS \wedge p1 \mapsto p2 \in CALLS$
  then $CALLS := CALLS - \{ p1 \mapsto p2 \}$ **end**;

Following J.R. Abrial [2,3], we may add others events to the current set of
events and we have to check two laws :
Law 1: *Our system remains "live" and the deadlock freedom property is a
safety property of our system. Formally, we have to state that the disjunction
of the guards of events must be always true and we have to generate proof
obligations, which are not automatically produced by the current tool.*
The following property is a safety property required by every abstract model:

$$\exists\, e\,.(e\,\in\,Events \land guard(e))$$

The condition for the abstract model $BASE_1$ is reduced to:

$$\exists (p1,p2).(\ p1,p2 \in PERSONS \land (p1 \mapsto p2 \notin CALLS \lor p1 \mapsto p2 \in CALLS\ )$$

Law 2: *We assume a fairness assumption over events; the set of new events can ever "take control" for ever.*
Fairness assumptions are critical, when one wants to ensure that a progress is eventual; J.-R. Abrial and L. Mussat [3] develop an extension of the B method to take into account dynamic properties such as eventuality properties and they introduce explicit counters for ensuring the fairness of the resulting systems. Their solution is very close to the proof rules of Apt and Olderog [4], where scheduling variables were used to keep only fair traces of systems. We do not address the question of fairness in our current report and leave it for further researches; predicate diagrams [10] can be used to state fairness assumptions and to generate proof obligations for fairness when refining models. Now, we use rules for transforming abstract models into more concrete ones. Two main transformations are applied on event systems: adding a new event and modifying an event.
Remember that we have to preserve safety properties and the invariant when applying these transformations, when it is possible. According to the UNITY approach [11], we will call them superpositions, since we superpose a new computation on an old one. When one applies a transformation rule, one has to check proof obligations stating the refinement relationship. If the new model introduces a new variable x, then a new event on x will refine a skip statement and it is clear that, the new event was observed such as a skip statement in the abstract model. Hence, no new proof obligation is required; when one strengthens a guard, new proof obligations are generated for stating the preservation of invariant:

- $[G_1 \Rightarrow S_1] \sqsubseteq [G_2 \Rightarrow S_2]$ where G_1 and G_2 are two conditions over system variables and they satisfy the following property : $G_2 \Rightarrow G_1$ and S_1 and S_2 are two substitutions over system variables which satisfy $S_1 \sqsubseteq S_2$ (S_2 refines S_1).
- $[S_1 \sqsubseteq S_1 \parallel S_2]$ where S_1 and S_2 are two substitutions over system variables which do not share variables.

We refine our current version of $BASE$, called $BASE_1$ into $BASE_2$; $BASE_2$ improves the view of our system, since it is clear that a call is possible, when the switch has authorised it. We do not know how authorisations are assigned but we use a variable to control and to record authorisations:

$$SYS\_AUTH \in PERSONS \leftrightarrow PERSONS$$

Jim is authorised to call Jane, when $Jim \mapsto Jane$ is in SYS_AUTH. We require that, when a user is authorised to call somebody else, he/she will loose the authorisation at the call creation; the authorisation should be checked before each call: $CALLS \cap SYS\_AUTH = \emptyset$.

Two new events are appended to our current system:

$System\_Adding\_Auth =$
  any $p1, p2$ **where** $p1 \in PERSONS \wedge p2 \in PERSONS$
$$\wedge \ p1 \mapsto p2 \notin CALLS$$
  then $SYS\_AUTH := SYS\_AUTH \cup \{ p1 \mapsto p2 \}$
  end;

$System\_Removing\_Auth =$
  any $p1, p2$ **where**
     $p1 \in PERSONS \wedge p2 \in PERSONS \wedge p1 \mapsto p2 \in SYS\_AUTH$
  then $SYS\_AUTH := SYS\_AUTH - \{p1 \mapsto p2 \}$
  end;

and we modify the event $Creating\_a\_call$ by strengthening the guard and reinforcing the computation:

$Creating\_a\_call =$
  any $p1, \ p2$ **where**
   $p1 \in PERSONS \wedge p2 \in PERSONS \wedge p1 \mapsto p2 \in SYS\_AUTH$
  then
     $CALLS := CALLS \cup \{ p1 \mapsto p2 \}$
   $\| \ SYS\_AUTH := SYS\_AUTH - \{ p1 \mapsto p2 \}$
  end;

The new invariant is trivially checked. The new model is $BASE_2$ and it refines $BASE_1$:

$$BASE_1 \longrightarrow BASE_2$$

The unfolding of abstract events and abstract traces continues. In $BASE_2$, when A is calling B, first A has to be authorised to call B and secondly the authorisation must be maintained. Removing an authorisation may be due to a decision of the switch system, which controls the call process: billing problem, off-hook, problem of the line ... We observe that an authorisation is a reply to a request of a user for calling somebody. Hence, we define a new model $BASE_3$. A new variable $USER\_REQ\_AUTH$ records the fact that a user A wants to call B:

$$USER\_REQ\_AUTH \in PERSONS \leftrightarrow PERSONS$$

New events are added and we modify the event which assigns an authorisation, since we know that it replies to a request.

$Creating\_a\_call$ =
  any $p1$, $p2$ **where**
   $p1 \in PERSONS \wedge p2 \in PERSONS \wedge p1 \mapsto p2 \in SYS\_AUTH$
  then
   $CALLS$:= $CALLS \cup \{\, p1 \mapsto p2 \,\}$
    $\|$ SYS_AUTH := $SYS\_AUTH - \{\, p1 \mapsto p2 \,\}$
  end;
$Halting\_a\_call$ =
  any $p1$, $p2$ **where**
   $p1 \in PERSONS \wedge p2 \in PERSONS \wedge p1 \mapsto p2 \in CALLS$
  then $CALLS$:= $CALLS - \{\, p1 \mapsto p2 \,\}$
  end;
$System\_Adding\_Auth$ =
  any $p1, p2$ **where**
   $p1 \in PERSONS \wedge p2 \in PERSONS \wedge p1 \mapsto p2 \notin (SYS\_AUTH \cup CALLS)$
            $\wedge$ $p1 \mapsto p2 \in USER\_REQ\_AUTH$
  then
   SYS_AUTH := $SYS\_AUTH \cup \{\, p1 \mapsto p2 \,\}$
    $\|$ $USER\_REQ\_AUTH$:= $USER\_REQ\_AUTH - \{\, p1 \mapsto p2 \,\}$
  end;
$System\_Refusing\_Auth$ =
  any $p1, p2$ **where**
   $p1 \in PERSONS \wedge p2 \in PERSONS \wedge p1 \mapsto p2 \in USER\_REQ\_AUTH$
  then
   $USER\_REQ\_AUTH$:= $USER\_REQ\_AUTH - \{\, p1 \mapsto p2 \,\}$
  end;
$System\_Removing\_Auth$ =
  any $p1, p2$ **where**
   $p1 \in PERSONS \wedge p2 \in PERSONS \wedge p1 \mapsto p2 \in SYS\_AUTH$
  then
   SYS_AUTH := $SYS\_AUTH - \{p1 \mapsto p2 \,\}$
  end;
$User\_Requesting\_Call$ =
  any $p1, p2$ **where**
   $p1 \in PERSONS \wedge p2 \in PERSONS$
  then
   $USER\_REQ\_AUTH$:= $USER\_REQ\_AUTH \cup \{\, p1 \mapsto p2 \,\}$
  end;
$User\_Halting\_Requesting\_Call$ =
  any $p1, p2$ **where**
   $p1 \in PERSONS \wedge p2 \in PERSONS \wedge p1 \mapsto p2 \in USER\_REQ\_AUTH$
  then
   $USER\_REQ\_AUTH$:= $USER\_REQ\_AUTH - \{p1 \mapsto p2 \,\}$
  end;

Fig. 3. Events of the abstract system $BASE_3$

VARIABLES *CALLS, SYS_AUTH, USER_REQ_AUTH*
INVARIANT
$\quad\quad$ *CALLS* $\in$ *PERSONS* $\leftrightarrow$ *PERSONS*
$\quad$ $\wedge$ *SYS_AUTH* $\in$ *PERSONS* $\leftrightarrow$ *PERSONS*
$\quad$ $\wedge$ *USER_REQ_AUTH* $\in$ *PERSONS* $\leftrightarrow$ *PERSONS*
$\quad$ $\wedge$ *CALLS* $\cap$ *SYS_AUTH* $=$ $\emptyset$
INITIALISATION *CALLS, SYS_AUTH, USER_REQ_AUTH* $:=$ $\emptyset, \emptyset, \emptyset$

Fig. 4. Invariant of the abstract system $BASE_3$

Following the process, we add more and more events and we make our abstract models very close to the required behavior of classical calls. Four other models are incrementally defined and checked by Atelier B. We have to check new proof obligations generated by the law 1 and we assign an abstract fairness to guarantee eventuality properties. Eventuality properties are out of the scope of the current paper. The final model $BASE_6$ looks like a very concrete model of POTS:

$$BASE_1 \rightarrow BASE_2 \rightarrow BASE_3 \rightarrow BASE_4 \rightarrow BASE_5 \rightarrow BASE_6$$

We have to address the problem of services, features and possible interactions by refining $BASE_1$, ..., $BASE_6$ models. Events systems are useful for understanding how services are working and what are safety properties. Hence, the communication with the customer/user involves a simple way to write safety requirements. We can use the ASSERTIONS clause and hide details of events. The invariant of $BASE_6$ tells us that a user A can not be in communication with himself/herself; the property is clearly accepted by the customer but the property was not true in the first abstract models.. We have strengthened the guard of operation *SYS_AUTH* by stating that p1 and p2 are distinct. Hence, the invariant has been incrementally defined and is in fact inductive. The invariant of $BASE_6$ expresses classical requirements of the basic phone system, as for instance *p1 can only communicate with only one person p2 at a time* (*CALLS* and *SYS_AUTH* are two partial into mappings; *USER_REQ_AUTH* is a partial mapping, because p1 and p2 may want to call concurrently p3). See [9] for more details.

$\quad\quad$ *CALLS* $\in$ *PERSONS* $\rightarrowtail$ *PERSONS*
$\quad$ $\wedge$ DOM(*CALLS*) $\cap$ RAN(*CALLS*) $=$ $\emptyset$
$\quad$ $\wedge$ *SYS_AUTH* $\in$ *PERSONS* $\rightarrowtail$ *PERSONS*
$\quad$ $\wedge$ *USER_REQ_AUTH* $\in$ *PERSONS* $\rightarrow$ *PERSONS*

System_Halting_Requesting_Call is an event which models the following fact: a person wants to call somebody and after a while, he/she decides to hang up. *System_Removing_Auth* is an event which models that the system does not give the authorisation ($p-2$ is busy or p_1 does not pay the bill). We have to remove the authorisation and this requirement was detected, when we modelled the event of hanging up. Finally, any person can not call two persons at the same time, since the following safety property is derived from the invariant in the ASSERTIONS clause:

$$CALLS \cap \text{ID}(PERSONS) = \emptyset$$

The refinement guarantees that no undesired behavior is possible. In our previous works [27,9] on using B for services, we were addressing state properties, we have composed invariants ie conjunction of invariants and we used the incremental development of abstract models called views. We add more and more details into our abstract models and we may want to add billing features.

3   Integration of services

Our method for integrating services is based on the analysis of generated proof obligations; we demonstrate the effectiveness of our method by showing how we have written services in our first work [9]. We consider three services TCS Terminating Screen Service, Originating Call Service and CF Call Forwarding. The main point is to keep explanations of choices done during the development; a proof obligation is obviously a very formal statement for explaining our choices and we have to provide with the customer an explanation in his/her language.

3.1   Two simple case studies TCS and OCS

Let us consider the first machine $BASE_1$. The integration of TCS (or OCS) is simple: we should strengthen the invariant of the machine by adding properties relating $CALLS$ and the new variable $TCSLIST$. The new variable $TCSLIST$ is a set of pairs m, n and if $m, n \in TCSLIST$, then m has forbidden n to call him. Then, the invariant states that $CALLS$ and $TCSLIST^{-1}$ have no common elements: this is the requirement of the service TCS. We denote $REL\_PERSONS \equiv PERSONS \leftrightarrow PERSONS$

INVARIANT
$$CALLS \in REL\_PERSONS$$
$$\land \ TCSLIST \in REL\_PERSONS \ \land \ OCSLIST \in REL\_PERSONS$$
$$\land \ CALLS \cap TCSLIST^{-1} = \emptyset \quad \land \ CALLS \cap OCSLIST = \emptyset$$

OCS is integrated in the same way but we add a condition over $OCSLIST$ ($CALLS \cap OCSLIST = \emptyset$) . Now, we generate new proof obligations and only one proof obligation remained unproved and is unprovable.

$not(PERSONS = \emptyset) \wedge PERSONS \in FIN(\mathbb{Z}) \wedge$
$CALLS\$1 \in REL\_PERSONS \wedge TCSLIST\$1 \in REL\_PERSONS \wedge$
$CALLS\$1 \cap TCSLIST\$1^{-1} = \emptyset \wedge SYS\_AUTH\$1 \in REL\_PERSONS \wedge$
$SYS\_AUTH\$1 \cap CALLS\$1 = \emptyset \wedge SYS\_AUTH\$1 \cap TCSLIST\$1^{-1} = \emptyset \wedge$
$TCSLIST = TCSLIST\$1 \wedge CALLS = CALLS\$1 \wedge$
$p1 \in PERSONS \wedge p2 \in PERSONS \wedge not(p1 \mapsto p2 \in CALLS\$1) \wedge$
$\Rightarrow$
$SYS\_AUTH\$1 \cup \{p1 \mapsto p2\} \cap TCSLIST\$1^{-1} = \emptyset$

The property is provable, when one adds the assumption $not(p1 \mapsto p2 : TCSLIST\$1^{-1})$. Hence, we have to strengthen the guard of the event *Creating_a_call*, when one integrates TCS and OCS: it is the only solution.

Creating_a_call $=$
  any $p1$, $p2$
  where
      $p1 \in PERSONS \wedge p2 \in PERSONS$
      $\wedge\ p1 \mapsto p2 \notin (CALLS \cup TCSLIST^{-1} \cup OCSLIST)$
  then $CALLS := CALLS \cup \{ p1 \mapsto p2 \}$ **end**;

Now, we have to integrate TCS and OCS into $BASE_2$ and we instantiate the diagram of the figure 1 with $BASE_1$, $BASE_2$ and TCS_1; we obtain an unproved proof obligation which is still unprovable. It looks like the previous one. The invariant of the new system $BASE_2$ is then:

$CALLS \in REL\_PERSONS \wedge SYS\_AUTH \in REL\_PERSONS$
$\wedge\ TCSLIST \in REL\_PERSONS \wedge OCSLIST \in REL\_PERSONS$
$\wedge\ CALLS \cap SYS\_AUTH = \emptyset \wedge CALLS \cap TCSLIST^{-1} = \emptyset$
$\wedge\ TCSLIST^{-1} \cap SYS\_AUTH = \emptyset$
$\wedge\ CALLS \cap OCSLIST = \emptyset \wedge OCSLIST \cap SYS\_AUTH = \emptyset$

The second refined view $BASE_2$ has two events, which can be used to control calls: *Creating_a_call* or *System_Adding_Auth*. *System_Adding_Auth* controls the authorization and both events provide a guard that can be strengthened to ensure the invariant; both transformations are possible but it is better to strengthen the guard of *System_Adding_Auth*, since its rôle is to control the authorizations.

System_Adding_Auth $=$
any $p1, p2$
  where
      $p1 \in PERSONS \wedge p2 \in PERSONS$
      $\wedge\ p1 \mapsto p2 \notin (CALLS \cup TCSLIST^{-1} \cup OCSLIST)$
  then $SYS\_AUTH := SYS\_AUTH \cup \{ p1 \mapsto p2 \}$
  end;

The next integration of TCS and OCS for the view $BASE_3$ leads us to several choices: either strengthening the guard of *System_Adding_Auth*, or strengthening the guard of *User_Requesting_Call*. If we choose to strengthen the guard of *User_Requesting_Call* for TCS, it means that the user knows the TCS list of the callee. It is not a good idea, since the callee's TCS list is probably locally managed. However, his calls can be filtered by his phone according to the OCS list. We obtain two choices TCS and OCS.

Strengthening the guard of *System_Adding_Auth* The new invariant states that the system can not authorize a call filtered by OCSLIST.

INVARIANT

$CALLS \in REL\_PERSONS \land SYS\_AUTH \in REL\_PERSONS$
$\land\ USER\_REQ\_AUTH \in REL\_PERSONS \land TCSLIST \in REL\_PERSONS$
$\land\ OCSLIST \in REL\_PERSONS \land CALLS \cap SYS\_AUTH = \emptyset$
$\land\ CALLS \cap TCSLIST^{-1} = \emptyset \land CALLS \cap OCSLIST = \emptyset$
$\land\ OCSLIST \cap SYS\_AUTH = \emptyset \land TCSLIST^{-1} \cap SYS\_AUTH = \emptyset$

The tool provides us two unprovable proof obligations from the analysis diagram; both are for the event *System_Adding_Auth*. The first one tells us that outgoing calls must be filtered by OCS.

$not(PERSONS = \emptyset) \land PERSONS \in FIN(\mathbb{Z}) \land$
$CALLS\$1 \in REL\_PERSONS \land SYS\_AUTH\$1 \in REL\_PERSONS \land$
$USER\_REQ\_AUTH\$1 \in REL\_PERSONS \land$
$TCSLIST\$1 \in REL\_PERSONS \land OCSLIST\$1 \in REL\_PERSONS \land$
$CALLS\$1 \cap SYS\_AUTH\$1 = \emptyset \land CALLS\$1 \cap TCSLIST\$1^{-1} = \emptyset \land$
$CALLS\$1 \cap OCSLIST\$1 = \emptyset \land OCSLIST\$1 \cap SYS\_AUTH\$1 = \emptyset \land$
$TCSLIST\$1^{-1} \cap SYS\_AUTH\$1 = \emptyset \land CALLS = CALLS\$1 \land$
$USER\_REQ\_AUTH = USER\_REQ\_AUTH\$1 \land$
$SYS\_AUTH = SYS\_AUTH\$1 \land p1 \in PERSONS \land p2 \in PERSONS \land$
$not(p1 \mapsto p2 \in SYS\_AUTH\$1) \land not(p1 \mapsto p2 \in CALLS\$1) \land$
$p1 \mapsto p2 \in USER\_REQ\_AUTH\$1 \land$
$\Rightarrow$
$not(p1 \mapsto p2 \in OCSLIST\$1)$

The second one recalls us that the system can not authorize calls that are forbidden by TCS.

$not(PERSONS = \emptyset) \land PERSONS \in FIN(\mathbb{Z}) \land$
$CALLS\$1 \in REL\_PERSONS \land SYS\_AUTH\$1 \in REL\_PERSONS \land$
$USER\_REQ\_AUTH\$1 \in REL\_PERSONS \land$
$TCSLIST\$1 \in REL\_PERSONS \land OCSLIST\$1 \in REL\_PERSONS \land$
$CALLS\$1 \cap SYS\_AUTH\$1 = \emptyset \land CALLS\$1 \cap TCSLIST\$1^{-1} = \emptyset \land$

$CALLS\$1 \cap OCSLIST\$1 \ = \ \emptyset \wedge \ OCSLIST\$1 \cap SYS\_AUTH\$1 \ = \ \emptyset \wedge$
$TCSLIST\$1^{-1} \cap SYS\_AUTH\$1 \ = \ \emptyset \wedge SYS\_AUTH \ = \ SYS\_AUTH\$1 \wedge$
$USER\_REQ\_AUTH = USER\_REQ\_AUTH\$1 \wedge CALLS \ = \ CALLS\$1 \wedge$
$p1 \in \ PERSONS \wedge p2 \in \ PERSONS \wedge not(p1 \mapsto p2 \in \ SYS\_AUTH\$1) \wedge$
$not(p1 \mapsto p2 \in \ CALLS\$1) \wedge p1 \mapsto p2 \in \ USER\_REQ\_AUTH\$1 \wedge$
$\Rightarrow$
$not(p1 \mapsto p2 \in \ TCSLIST\$1^{-1})$

The new view has a strengthened event *System_Adding_Auth*; the guard of *System_Adding_Auth* is strengthened as follows:

System_Adding_Auth =
  any $p1, p2$ **where**
    $p1 \in \ PERSONS \wedge p2 \in \ PERSONS$
    $\wedge \ p1 \ \mapsto \ p2 \notin \ (SYS\_AUTH \cup CALLS \cup TCSLIST^{-1} \cup OCSLIST)$
    $\wedge \ p1 \ \mapsto \ p2 \in \ USER\_REQ\_AUTH\_TO\_SYS$
  then
      $SYS\_AUTH \ := \ SYS\_AUTH \cup \{ p1 \ \mapsto \ p2 \}$
    $\| \ USER\_REQ\_AUTH \ := \ USER\_REQ\_AUTH - \{p1 \ \mapsto \ p2\}$
  end;

Strengthening the guard of *User_Requesting_Call* The choice corresponds to a new invariant that satisfies the new idea. The requirement on OCSLIST is that if p is not authorised to call p, then p can not request an authorization to the system. The control is local and two proof obligations remain unprovable.

INVARIANT
  $CALLS \in \ REL\_PERSONS \wedge SYS\_AUTH \in \ REL\_PERSONS$
$\wedge \ USER\_REQ\_AUTH \in \ REL\_PERSONS$
$\wedge \ TCSLIST \in \ REL\_PERSONS \wedge OCSLIST \in \ REL\_PERSONS$
$\wedge \ CALLS \ \cap SYS\_AUTH = \ \emptyset \wedge CALLS \cap \ TCSLIST^{-1} \ = \ \emptyset$
$\wedge \ CALLS \ \cap OCSLIST \ = \ \emptyset \wedge OCSLIST \cap \ SYS\_AUTH \ = \ \emptyset$
$\wedge \ TCSLIST^{-1} \cap SYS\_AUTH = \emptyset \wedge OCSLIST \cap USER\_REQ\_AUTH = \emptyset$

We have to keep the test on TCSLIST in the event *System_Adding_Auth*, but the next proof obligation tell us to strengthen the guard of the event *USER_REQ_AUTH* to fulfill the requirement on *OCSLIST*.

$not(PERSONS \ = \ \emptyset) \wedge PERSONS \in \ FIN(\mathbb{Z}) \wedge$
$CALLS\$1 \in \ REL\_PERSONS \wedge SYS\_AUTH\$1 \in \ REL\_PERSONS \wedge$
$USER\_REQ\_AUTH\$1 \in \ REL\_PERSONS \wedge CALLS \ = \ CALLS\$1$
$TCSLIST\$1 \in \ REL\_PERSONS \wedge OCSLIST\$1 \in \ REL\_PERSONS \wedge$
$CALLS\$1 \cap SYS\_AUTH\$1 \ = \ \emptyset \wedge CALLS\$1 \cap TCSLIST\$1^{-1} \ = \ \emptyset \wedge$
$CALLS\$1 \cap OCSLIST\$1 \ = \ \emptyset \wedge OCSLIST\$1 \cap SYS\_AUTH\$1 \ = \ \emptyset \wedge$
$TCSLIST\$1^{-1} \cap SYS\_AUTH\$1 = \emptyset \wedge p1 \in \ PERSONS \wedge$
$OCSLIST\$1 \cap USER\_REQ\_AUTH\$1 = \emptyset \wedge p2 \in \ PERSONS \wedge$

$$USER\_REQ\_AUTH = USER\_REQ\_AUTH\$1 \wedge$$
$$SYS\_AUTH = SYS\_AUTH\$1 \wedge$$
$$\Rightarrow$$
$$not(p1 \mapsto p2 \in OCSLIST\$1)$$

Finally, we obtain the following refined events for the new view integrating OCS and TCS into $BASE_2$ and every proof obligation is obviously proved.

$User\_Requesting\_Call =$
    any $p1, p2$ **where**
        $p1 \in PERSONS \wedge p2 \in PERSONS$
    $\wedge\ p1 \mapsto p2 \notin OCSLIST$
    then
        $USER\_REQ\_AUTH := USER\_REQ\_AUTH \cup \{p1 \mapsto p2\}$
    end;

$System\_Adding\_Auth =$
    any $p1, p2$ **where**
        $p1 \mapsto p2 \in USER\_REQ\_AUTH$
    $\wedge\ p1 \mapsto p2 \notin (SYS\_AUTH \cup CALLS \cup TCSLIST^{-1})$
    then
        $SYS\_AUTH := SYS\_AUTH \cup \{ p1 \mapsto p2 \}$
        $\|\ USER\_REQ\_AUTH := USER\_REQ\_AUTH - \{p1 \mapsto p2\}$
    end;

3.2 Updating the TCS list

It is easy to add new events to our current views; we notice that works on features do not mention events that can modify the list $TCSLIST$. In fact, it may be a source of interactions. We add the event $UpDating\_TCS$ to our view $BASE_3TCS_3$ without toil but without care.

$UpDating\_TCS =$ **any** $pp, tcspp$ **where**
        $pp \in PERSONS \wedge tcspp \subseteq PERSONS$
    then
        $TCSLIST := TCSLIST \rhd \{pp\} \cup tcspp \times \{pp\}$
    end;

The new event $UpDating\_TCS$ modifies the set of persons who are forbidden to be called by pp; the event updates the list TCSLIST by removing every pair qq,pp in TCSLIST and add a new set of pairs $tcspp \times \{pp\}$ where $tcspp$ is the set of persons who decide to add pp on their TCS list. Without toil and without care leads to two unprovable proof obligations!

$not(PERSONS = \emptyset) \wedge PERSONS \in FIN(\mathbb{Z}) \wedge$
$CALLS\$1 \in REL\_PERSONS \wedge SYS\_AUTH\$1 \in REL\_PERSONS \wedge$
$USER\_REQ\_AUTH\$1 \in REL\_PERSONS \wedge$
$CALLS\$1 \cap SYS\_AUTH\$1 = \emptyset \wedge TCSLIST\$1 \in REL\_PERSONS \wedge$
$OCSLIST\$1 \in REL\_PERSONS \wedge CALLS\$1 \cap TCSLIST\$1^{-1} = \emptyset \wedge$
$CALLS\$1 \cap OCSLIST\$1 = \emptyset \wedge OCSLIST\$1 \cap SYS\_AUTH\$1 = \emptyset \wedge$
$TCSLIST\$1^{-1} \cap SYS\_AUTH\$1 = \emptyset \wedge$
$OCSLIST\$1 \cap USER\_REQ\_AUTH\$1 = \emptyset \wedge$
$USER\_REQ\_AUTH = USER\_REQ\_AUTH\$1 \wedge$
$SYS\_AUTH = SYS\_AUTH\$1 \wedge CALLS = CALLS\$1 \wedge$
$pp \in PERSONS \wedge tcspp \subseteq PERSONS \wedge$
$\Rightarrow$
$CALLS\$1 \cap (TCSLIST\$1 \rhd \{pp\} \cup tcspp \times \{pp\})^{-1} = \emptyset$

The first unprovable proof obligation tells us that one can not authorize a person to update a TCS list, while the person is called by pp; we can solve it by adding a condition over pp in the guard. We check first if $pp \notin \text{DOM}(CALLS)$ (pp is not calling somebody-else) . The prover successfully proves the new generated proof obligation. However, there is still one unprovable proof obligation for the following event which models the updating of TCS subscribers with respect to pp.

$UpDating\_TCS =$
    any pp, $tcspp$ **where**
        $pp \in PERSONS \wedge pp \notin \text{DOM}(CALLS) \wedge tcspp \subseteq PERSONS$
    then
        $TCSLIST := TCSLIST \rhd \{pp\} \cup tcspp \times \{pp\}$
    end;

The last unprovable proof obligation states that we have to check if pp has been already authorised to call somebody. It may lead to a state which does not satisfy by the invariant. In fact, the following proof obligation tells us simply to strengthen more strongly the guard by adding $pp \notin \text{DOM}(SYS\_AUTH)$ to the current guard.

$not(PERSONS = \emptyset) \wedge PERSONS \in FIN(\mathbb{Z}) \wedge$
$CALLS\$1 \in REL\_PERSONS \wedge SYS\_AUTH\$1 \in REL\_PERSONS \wedge$
$USER\_REQ\_AUTH\$1 \in REL\_PERSONS \wedge$
$CALLS\$1 \cap SYS\_AUTH\$1 = \emptyset \wedge TCSLIST\$1 \in REL\_PERSONS \wedge$
$OCSLIST\$1 \in REL\_PERSONS \wedge CALLS\$1 \cap TCSLIST\$1^{-1} = \emptyset \wedge$
$CALLS\$1 \cap OCSLIST\$1 = \emptyset \wedge OCSLIST\$1 \cap SYS\_AUTH\$1 = \emptyset \wedge$
$TCSLIST\$1^{-1} \cap SYS\_AUTH\$1 = \emptyset \wedge$
$OCSLIST\$1 \cap USER\_REQ\_AUTH\$1 = \emptyset \wedge$
$USER\_REQ\_AUTH = USER\_REQ\_AUTH\$1 \wedge$
$SYS\_AUTH = SYS\_AUTH\$1 \wedge CALLS = CALLS\$1 \wedge$
$pp \in PERSONS \wedge tcspp \subseteq PERSONS \wedge$
$\Rightarrow$
$(TCSLIST\$1 \rhd \{pp\} \cup tcspp \times \{pp\})^{-1} \cap SYS\_AUTH\$1 = \emptyset$

The new event is then checked by the prover and no more unprovable proof obligation is generated.

$UpDating\_TCS$ =
   any pp, $tcspp$ **where**
         $pp \in PERSONS \wedge pp \notin$ DOM $(CALLS)$
    $\wedge\ pp \notin$ DOM$(SYS\_AUTH) \wedge tcspp \subseteq PERSONS$
   then
       $TCSLIST\ :=\ TCSLIST \vartriangleright \{pp\} \cup (tcspp \times \{pp\})$
   end

We have done a choice during the development and the choice has to be confirmed by the customer. The customer is the provider who will develop the service. We do not give the proofs of our views to the customer but we give a list of proposals and model the behaviour selected by our customer.

3.3   Modelling CF Call Forwarding

Call Forwarding allows the user to decide to have incoming calls diverted to an alternative user/phone. Call Forwarding performs a translation, changing the called number into the actual destination line. Call Forwarding has a variety of forms and names as Call Diversion, Call Forwarding on Busy Line, Call Forwarding Don't Answer, Call Rerouting Distribution, Follow-me Diversion. Six refined models are defined from the six views of $BASE$ and we introduce the following variables:

- CF^* is a partial function over PERSONS and it models the partial closure of the transfer function, when no cycle is detected. A cycle may exist, when a person p_1 transfer to p_2, p_2 transfers to p_3, ..., p_{n-1} transfers to p_n and p_n transfers to p_1. Usually, the transfer function is updated by the subscriber of CF but we add a new event which models the updating of the call forward..
- $CF\_SUBSCRIBER$ is a set of persons, who have subscribed Call Forwarding.

The refinement is proved by the tool Atelier B. The new step is to refine $BASE_2$ and we obtain a refinement of $BASE_2$ into $BASE_2 \oplus CF_2$ and a refinement of $BASE_1 \oplus CF_1$ into $BASE_2 \oplus CF_2$. Finally, no interaction is detected. Now, we refine $BASE_3$ into $BASE_3 \oplus CF_3$ and, after the completion of the proof process, a proof obligation remains unproved.

INVARIANT
     $CALLS \in PERSONS \leftrightarrow PERSONS$
   $\wedge\ CF^* \in PERSONS \nrightarrow PERSONS$
   $\wedge\ CF\_SUBSCRIBER \subseteq PERSONS$
   $\wedge\ \forall pp.(pp \in PERSONS - CF\_SUBSCRIBER \Rightarrow CF^*(pp) = pp)$
   $\wedge\ \forall pp.(pp \in$ RAN$(CF^*) \Rightarrow CF^*(pp) = pp)$

The integration of CF into $BASE$ raises the following question: what is a good call, when one forwards it? The main problem of a forwarded call is to define the target of the call; it may appear that a cycle exists through different forwarded calls. Hence, the problem is to determine $CF^*(p2)$, when $p1$ is calling $p2$, and, when $p2$ has subscribed the CF service. If $p2$ is not in the domain of CF^*, it means that there is a cycle because of the different forwarded calls. Since we were testing on the pair $p1 \mapsto p2$, now we have to test on the pair $p1 \mapsto CF^*(p2)$. However, we modify the view of the customer; $p1$ wants to call $p2$ and not necessarily $CF^*(p2)$. We decide to let unmodified the previous guard and we obtain the following event $Creating\_a\_call$ in the view $BASE1$:

$Creating\_a\_call =$
    any $p1$, $p2$ **where**
        $p1 \in PERSONS \wedge p2 \in PERSONS \wedge p2 \in \text{DOM}(CF^*)$
        $\wedge\ p1 \mapsto p2 \notin CALLS \wedge p1 \mapsto CF^*(p2) \notin CALLS$
    then
        $CALLS := CALLS \cup \{ p1 \mapsto CF^*(p2) \}$
    end;

The updating of the CF data is simple and we avoid to make complex the event; we suppose that it preserves the invariant. We notice that the updating of CF may modify the transitive closure of CF.

$UpDating\_CF =$ **begin**
    $CF^* :\in \{ff \mid ff \in PERSONS \rightarrowtail PERSONS$
        $\wedge\ \forall pp.(pp \in PERSONS-CF\_SUBSCRIBER \Rightarrow ff(pp) = pp)$
        $\wedge\ \forall pp.(pp \in \text{RAN}(ff) \Rightarrow ff(pp) = pp) \}$
end

The first view of the integration of CF into $BASE$ is now completed; the integration of the event $UpDating\_CF$ leads us to a problem , when refining $BASE_1 \oplus CF_1$. $BASE_2$ is refined into $BASE_2 \oplus CF_2$, but an unprovable proof obligation is generated, when one wants to refine $BASE_1 \oplus CF_1$ into $BASE_2 \oplus CF_2$.

$p1 \in PERSONS \wedge p2 \in PERSONS \wedge$
$p1 \mapsto p2 \in SYS\_AUTH\$1$
$\Rightarrow$
$\exists(p1\$0, p2\$0).(p1\$0 \in PERSONS \wedge p2\$0 \in PERSONS$
        $\wedge\ p2\$0 \in \text{DOM}(CF^*\$1)$
        $\wedge\ not(p1\$0 \mapsto p2\$0 \in CALLS\$1)$
        $\wedge\ not(p1\$0 \mapsto CF^*\$1(p2\$0) \in CALLS\$1)$
        $\wedge\ CALLS\$1 \cup \{p1 \mapsto p2\} = CALLS\$1 \cup \{p1\$0 \mapsto CF^*\$1(p2\$0)\})$

The unprovable proof obligation might be provable, if there exists $p3$ such that $p2 = CF^*(p3)$. This is exactly what we can get by adding a pair to SYS_AUTH. However, as long as the pair is not yet accepted for the call, $p2$ can not change his call forward. We must strengthen the invariant and the event $UpDating\_CF$ by expressing that a person in the codomain of SYS_AUTH can not change his call forward. The discussion with the customer can confirm this choice, since if the phones of the codomain of SYS_AUTH are ringing in $BASE_6$, then they are busy and can not modify their call forwards.

$UpDating\_CF$ = **begin**
$\quad CF^{*} :\in \{ f\!f \mid f\!f \in PERSONS \rightarrowtail PERSONS$
$\qquad\qquad \wedge \ \forall pp.(pp \in PERSONS - CF\_SUBSCRIBER \Rightarrow f\!f(pp) = pp)$
$\qquad\qquad \wedge \ \forall pp.(pp \in \text{RAN}(f\!f) \Rightarrow f\!f(pp) = pp)$
$\qquad\qquad \wedge \ \forall pp.(pp \in \text{RAN}(SYS\_AUTH) \Rightarrow f\!f(pp) = pp) \ \}$
end

4 Concluding remarks and future works

The paper extends the work [9] on the B system approach called here the system-mate approach and it provides more details on the way to derive refined views from combined views of telecommunications services. The characterisation of a feature interaction is based on the unprovability of one or more proof obligations; the incremental process for writing an abstract model or a view makes its validation easier. We agree that it is probably an original approach that maybe appeared technically too difficult and we have emphasised the translation process between the system-mate (specifier) and the customer. It is also probably easier to use a model checker rather than a theorem prover, but our method is mainly used for analysing requirements of services in an *a priori* policy. A consequence of our work is that we are able to write a list of requirements and a list of views for services; our proposal is validated by a theorem prover and is a contract with the customer. The combination of both approaches model checking and theorem proving would be beneficial but the B approach integrates the specifier into the process of understanding what the customer really wants; when a model checker tells us that a bug is discovered in a formal complex specification, it is a hint for understanding where is the bug - scenarios are generated -, but it is too late and a rewriting of the abstract model has to be carried out, but by theorem proving and without technical limits. Our approach for services is new with respect to the integration of semantics into the development process by proving proof obligations. The analysis pattern is the key concept of our method and we can play with views and the non monotonicity of features integration. We have not taken into account all details related to the services modeling and to the architecture-related aspects. The B event-based approach is under development and is close to works on TLA/TLA^{+}[23] and to works on actions systems [29]; further researches are needed for exploring structuring mechanisms and methodological issues. Our abstract models for $BASE$ do not mention billing facilities and variables and events for controlling costs of calls and it is very useful to model costs when services related to billing are used. Hence, we have to refine our basic model beyond $BASE_6$ to manage billing and billing-related services such as RB (Reverse Billing).

References

1. J.-R. Abrial. *The B book - Assigning Programs to Meanings*. Cambridge University Press, 1996.
2. J.-R. Abrial. Extending B without changing it (for developing distributed systems). In H. Habrias, editor, *1st Conference on the B method*, pages 169–190, November 1996.
3. J.-R. Abrial and L. Mussat. Introducing dynamic constraints in B. In D. Bert, editor, *B'98 :Recent Advances in the Development and Use of the B Method*, volume 1393 of *Lecture Notes in Computer Science*. Springer-Verlag, 1998.
4. K. R. Apt and E. R. Olderog. Proof rules and transformations dealing with fairness. *Science of Computer Programming*, 3:65–100, 1983.
5. C. Areces, W. Bouma, and M. de Rijke. Feature interaction as a satisfiability problem. In M. Calder and E. Magill, editors, *Feature Interactions in Telecommunications and Software Systems VI*. IOS Press, 2000. In [8].
6. J. Blom, B. Johnsson, and L. Kempe. Automatic detection of feature interactions in temporal logic. In K. E. Cheng and T. Ohta, editors, *Feature Interactions in Telecommunications Systems*, pages 1–19. IOS Press, 1996. [12].
7. L. G. Bouma and H. Velthuijsen, editors. *Feature Interactions in Telecommunications Systems*. IOS Press, 1994.
8. M. Calder and E. Magill, editors. *Feature Interactions in Telecommunications and Software Systems VI*, Glasgow, 1998. IOS Press.
9. D. Cansell and D. Méry. Playing with abstraction and refinement for managing features interactions - a methodological approach to feature interaction problem. In J. P. Bowen, S. Dunne, A. Galloway, and S. King, editors, *ZB2000 Conference*, volume 1878 of *Lecture Notes in Computer Science*, York, August 2000. Springer Verlag.
10. D. Cansell, D. Méry, and S. Merz. Predicate diagrams for the verification of reactive systems. In A. Galloway and B. Stoddart, editors, *IFM2000*, Saarbrücken, November 2000. Springer Verlag.
11. K. M. Chandy and J. Misra. *Parallel Program Design A Foundation*. Addison-Wesley Publishing Company, 1988. ISBN 0-201-05866-9.
12. K. E. Cheng and T. Ohta, editors. *Feature Interactions in Telecommunications Systems*. IOS Press, 1996.
13. P. Combes and S. Pickin. Formalisation of a user view of network and services for feature interaction detection. In L. G. Bouma and H. Velthuijsen, editors, *Feature Interactions in Telecommunications Software System*, pages 120–135. IOS Press, 1994. [7].
14. P. Dini, R. Boutaba, and L. Logrippo, editors. *Feature Interactions in Telecommunications Newtworks IV*, Montreal, 1997. IOS Press.
15. L. du Bousquet, F. Ouebdessalam, J.-L. Richier, and N. Zuanon. Incremental Feature Validation: A Synchronous Point of View. In K. Kimbler and W. Bouma, editors, *Feature Interaction Workshop*. IOS Press, 1998. In [22].

16. M. Frappier, A. Mili, and J. Desharnais. Detecting Feature Interaction on Relational Specifications. In P. Dini, R. Boutaba, and L. Logrippo, editors, *Feature Interaction Workshop*. IOS Press, 1997. In [14].
17. A. Gammelgaard and J. E. Kristensen. Interaction detection, a logical approach. In L. G. Bouma and H. Velthuijsen, editors, *Feature Interactions in Telecommunications Systems*, pages 178–196. IOS Press, 1994.
18. J.-P. Gibson, G. Hamilton, and D. Méry. Integration problems in telephone feature requirements. In A. Galloway and K. Taguchi, editors, *IFM'99 Integrated Formal Methods 1999*, Workshop In Computing Science, YORK, June 1999. Springer Verlag.
19. IEEE, editor. *Special Section Managing Feature Interactions in Telecommunications Sofware Systems*, volume 24. IEEE Computer Society, October 1998.
20. M. Jackson and P. Zave. Distributed feature composition - a virtual architecture for telecommunications services. *IEEE Transactions on Software Engineering*, 24(10):831–847, October 1998.
21. D. O. Keck and P. J. Kuehn. The feature and service interaction problem in telecommunications systems: A survey. *IEEE Transactions on Software Engineering*, 24(10):779–796, October 1998. In [19].
22. K. Kimbler and L. G. Bouma, editors. *Feature Interactions in Telecommunications and Software Systems V*, Lund, 1998. IOS Press.
23. L. Lamport. A temporal logic of actions. *Transactions On Programming Languages and Systems*, 16(3):872–923, May 1994.
24. F. J. Lin, H. Liu, and A. Ghosh. A Methodology for Feature Interaction Detection in the AIN 0.1 Framework. *IEEE Transactions on Software Engineering*, 24(10):797 – 817, October 1998. In [19].
25. B. Mermet and D. Méry. Incremental specification of telecommunication services. In M. Hinchey, editor, *First IEEE International Conference on Formal Engineering Methods (ICFEM)*, Hiroshima, November 1997. IEEE.
26. B. Mermet and D. Méry. Safe combinations of services using B. In John McDermid, editor, *SAFECOMP97 The 16th International Conference on Computer Safety, Reliability and Security*, York, September 1997. Springer Verlag.
27. B. Mermet and D. Méry. Service specifications to B, or not to B. In Mark Ardis, editor, *Second Workshop on Formal Methods in Software Practice*, Clearwater Beach, Florida, March 4-5 1998. ACM Press.
28. M. Plath and M. Ryan. Plug-and-play features. In K. Kimbler and W. Bouma, editors, *Feature Interaction Workshop*. IOS Press, 1998. In [22].
29. E. Sekerinski and K. Sere, editors. *Program Development by Refinement*. Springer, 1999.
30. STERIA - Technologies de l'Information, Aix-en-Provence (F). *Atelier B, Manuel Utilisateur*, 1998. Version 3.5.
31. K. Turner. Validating Architectural Feature Descriptions using LOTOS. In K. Kimbler and W. Bouma, editors, *Feature Interaction Workshop*. IOS Press, 1998. In [22].
32. Union Internationale des Télécommunications. Introduction à l'ensemble de capacités 1 du réseau intelligent. Technical Report UIT-T Q.1211, Union Internationale des Télécommunications, March 1993. Réseau Intelligent.
33. T. Yoneda and T. Ohta. A Formal Approach for Definition and Detection of Feature Intercation. In K. Kimbler and W. Bouma, editors, *Feature Interaction Workshop*. IOS Press, 1998. In [22].

Proving feature non-interaction with Alternating-Time Temporal Logic

Franck Cassez[1], Mark Dermot Ryan[2], and Pierre-Yves Schobbens[3]

[1] IRCCyN – BP 92101, 1 rue de la Noë, 44321 Nantes cedex 03, France.
Franck.Cassez@ircyn.ec-nantes.fr
[2] School of Computer Science, University of Birmingham, Edgbaston,
Birmingham B15 2TT, England. mdr@cs.bham.ac.uk
[3] Institut d'Informatique, Facultés Universitaires de Namur, Rue
Grandgagnage 21, 5000 Namur, Belgium. pys@info.fundp.ac.be

1 Introduction

Feature Interaction. When engineers design a system with features, they wish
to have methods to prove that the features do not interact in ways which are
undesirable. A considerable literature is devoted to this 'feature interaction
problem' [13,5]. One approach to demonstrating that features do not interact
undesirably is to equip them with properties which are intended to hold
of a system having the feature [18]. In this view, a feature is a pair (F, ϕ)
consisting of the implementation of the feature F and a set of properties
ϕ. Integrating a feature (F, ϕ) with a base system S consists of modifying
the base system in the way described by the feature implementation and
obtaining $S + F$. The integration is deemed successful if the resulting system
satisfies the set of properties ϕ corresponding to the feature. Evidence that
a feature (F_1, ϕ_1) does not negatively interact with feature (F_2, ϕ_2) may be
obtained by verifying that introducing F_2 in $S + F_1$, (obtaining $S + F_1 + F_2$)
does not destroy the properties ϕ_1 previous introduced by feature F_1, and
vice versa.

Model-Checking. Model checking [10] may be used to show that the featured
system has the desired properties (cf. [6,18,17,15]). However, the approach
described may involve checking the same property again and again each time
a new feature is introduced, to check that a previously introduced feature
has not been broken. Since model checking is computationally expensive, it is
worthwhile to find methods which avoid these re-checks. For example, we may
be able to prove generally that a certain feature does not destroy properties
in a certain class. This would obviate the need to re-check properties in that
class when the feature is introduced.

Users' Viewpoints and Conservative Features. A general result defining a
class of properties which are provably not broken by the introduction of a
new feature could be inspired by a number of intuitions. In this paper, we
develop such a result based on the idea that a feature which *adds* to the

capabilities of a *user* (and does not subtract from them) should not break properties which *assert* capabilities of this user. Let us look at the notion of capabilities in more detail.

A telephone system is usually made of a number of users and a network managing the calls. Many features of telephone systems are designed so that they add to the capabilities (or *powers*), of the subscribing user, without subtracting from them. For example, a user j who subscribes to call-forwarding now has the power to forward his or her calls to another user, but has not lost any capabilities in the process. When this is the case, we say that the feature is j-conservative. More generally, if U is a set of users, a feature is U-conservative if any behaviour of the system which U could enforce before the feature was added, U can also enforce it after the feature is added. The principal idea in this paper is that a U-conservative feature does not break properties which assert capabilities of the group U of users.

Framework. To formalise this intuition, we propose to use Alternating-time Temporal Logic [4] (ATL), which allows us to describe precisely the properties of the different *agents* (or users) involved in a system, and the *strategies* they have for achieving their goals. ATL is a branching temporal logic based on game theory. It contains the usual temporal operators (next, always, until) plus cooperation modalities $\langle\!\langle A \rangle\!\rangle \phi$, where A is a set of players. This modality quantifies over the set of behaviours and means that A has a collective strategy to enforce ϕ, whatever the choices of the other players. ATL generalises CTL, and similarly ATL* generalises CTL*, μ-ATL generalises the μ-calculus. These logics can be model-checked by generalising the techniques of CTL, often with the same complexity.

Outline of the paper. In section 2 we recall the basic concepts of ATL and ATL* and their semantics on ATSs. This section also introduces the Mocha-like language we use to describe reactive systems. The next section 3 which is the core of the paper describes the feature construct for our Mocha-like language and states the properties-preserving theorem. Finally in section 4 we discuss some directions for future work.

2   Alternating-time temporal logic

Alternating-time temporal logic (ATL) is based on CTL. Let us first recall a few facts about CTL. CTL [9] is a branching-time temporal logic in which we can express properties of reactive systems. For example, properties of cache-coherence protocols [16], telephone systems [18], and communication protocols have been expressed in CTL. One problem with CTL is that it does not distinguish between different sources of non-determinism. In a telephone system, for example, the different sources include individual users, the environment, and internal non-determinism in the telephone exchange. CTL provides the A quantifier to talk about all paths, and the E quantifier to assert the existence of a path. $A\psi$ means that, no matter how the non-determinism is resolved, ψ will be true of the resulting path. $E\psi$ asserts that, for at least one way of resolving the non-determinism, ψ will hold. But because CTL does not distinguish between different types of non-determinism, the A quantifier is often too strong, and the E quantifier too weak. For example, if we want to say that *user i can converse with user j*, CTL allows us to write the formulas

$$\texttt{A}\Diamond\texttt{talking}(i,j), \quad \texttt{E}\Diamond\texttt{talking}(i,j).$$

The first one says that in all paths, somewhere along the path there is a state in which i is talking to j, and is clearly much stronger than the intention. The second formula says that there is a path along which i is eventually talking j. This formula is weaker than the intention, because to obtain that path we may have to make choices on behalf of all the components of the system that behave non-deterministically. What we wanted to say is that users i and j can resolve their non-deterministic choices in such a way that, no matter how the other users or the system or the environment behaves, all the resulting paths will eventually have a state in which i is talking j. Of course, the fact that i is talking to j requires the cooperation of j. This subtle differences in expressing the properties we want to check can be captured accurately with ATL.

Alternating-time temporal logic (ATL) [4] generalises CTL by introducing *agents*, which represent different sources of non-determinism. In ATL the A and E path quantifiers are replaced by a unique path quantifier $\langle\!\langle A \rangle\!\rangle$, indexed by a subset A of the set of agents. The formula $\langle\!\langle A \rangle\!\rangle\psi$ means that the agents in A can resolve their non-deterministic choices such that, no matter how the other agents resolve their choices, the resulting paths satisfy ψ. We can express the property that user i has the power, or capability, of talking to j by the ATL formula[1]

$$\langle\!\langle i \rangle\!\rangle\Diamond\texttt{talking}(i,j).$$

We read $\langle\!\langle A \rangle\!\rangle\psi$ as saying that the agents in A can, by cooperating together, force the system to execute a path satisfying ψ. If A is the empty set of agents, $\langle\!\langle A \rangle\!\rangle\psi$ says that the system will execute ψ without the cooperation of any agents at all; in other words, $\langle\!\langle \emptyset \rangle\!\rangle\psi$ is equivalent to $A\psi$ in CTL. Dually, $\langle\!\langle \Sigma \rangle\!\rangle\psi$ (where Σ is the entire set of agents) is a weak assertion, saying that if all the agents conspire together they may enforce ψ, which is equivalent to $E\psi$ in CTL.

2.1   ATL and ATL*

Let P be a set of atomic propositions and Σ a set of agents. The syntax of ATL is given by

$$\phi ::= p \mid \top \mid \neg\phi_1 \mid \phi_1 \vee \phi_2 \mid \langle\!\langle A \rangle\!\rangle[\phi_1 \text{ U } \phi_2] \mid \langle\!\langle A \rangle\!\rangle\Box\phi_1 \mid \langle\!\langle A \rangle\!\rangle\bigcirc\phi_1$$

---

[1] We write $\langle\!\langle i \rangle\!\rangle\psi$ instead of $\langle\!\langle \{i\} \rangle\!\rangle\psi$.

where $p \in P$ and $A \subseteq \Sigma$. We use the usual abbreviations for $\rightarrow$, $\wedge$ in terms of $\neg$, $\vee$. The operator $\langle\!\langle\ \rangle\!\rangle$ is a path quantifier, and $\bigcirc$ (*next*), $\square$ (*always*) and U (*until*) are temporal operators. The logic ATL is similar to the branching-time logic CTL, except that path quantifiers are parameterised by sets of agents. As in CTL, we write $\langle\!\langle A \rangle\!\rangle \Diamond \phi$ for $\langle\!\langle A \rangle\!\rangle [\top \cup \phi]$.

While the formula $\langle\!\langle A \rangle\!\rangle \psi$ means that the agents in A can cooperate to make ψ true (they can "enforce" ψ), the dual formula $[[A]]\psi$ means that the agents in A cannot cooperate to make ψ false (they cannot "avoid" ψ). The formulas $[[A]]\Diamond\phi$, $[[A]]\square\phi$, and $[[A]]\bigcirc\phi$ stand for $\neg\langle\!\langle A \rangle\!\rangle\square\neg\phi$, $\neg\langle\!\langle A \rangle\!\rangle\Diamond\neg\phi$, and $\neg\langle\!\langle A \rangle\!\rangle\bigcirc\neg\phi$.

The logic ATL* generalises ATL in the same way that CTL* generalises CTL, namely by allowing path quantifiers and temporal operators to be nested arbitrarily.

For a subset $A \subseteq \Sigma$ of agents, the fragment $\langle\!\langle A \rangle\!\rangle$-ATL of ATL consists of ATL formulas whose only modality is $\langle\!\langle A \rangle\!\rangle$, and that does not occur within the scope of a negation. The $\langle\!\langle A \rangle\!\rangle$-ATL* fragment of ATL* is defined similarly.

2.2 Alternating transitions systems

Whereas the semantics of CTL is given in terms of transition systems, the semantics of ATL is given in terms of *alternating transition systems* (ATSs). An ATS over a set of atomic propositions P and a set of agents Σ is a triple $S = (Q, \pi, \delta, I)$, where Q is a set of states and $\pi : Q \rightarrow 2^P$ maps each state to the set of propositions that are true in it, and

$$\delta : Q \times \Sigma \rightarrow 2^{2^Q}$$

is a transition function which maps a state and an agent to a non-empty set of *choices*, where each choice is a non-empty set of possible next states. If the system is in a state q, each agent a chooses a set $Q_a \in \delta(q, a)$; the system will move to a state which is in $\bigcap_{a \in \Sigma} Q_a$. We require that the system is non-blocking and that the agents together choose a unique next state; that is, for every q and every tuple $(Q_a)_{a \in \Sigma}$ of choices $Q_a \in \delta(q, a)$, we require that $\bigcap_{a \in \Sigma} Q_a$ is a singleton. Similarly, the initial state is specified by $I : \Sigma \rightarrow 2^{2^Q}$. I maps each agent to a set of choices. The agents together choose a single initial state: for each tuple $(Q_a)_{a \in \Sigma}$ of choices $Q_a \in I(a)$, we require that $\bigcap_{a \in \Sigma} Q_a$ is a singleton.

For two states q and q', we say that q' is a *successor* of q if, for each $a \in \Sigma$, there exists $Q' \in \delta(q, a)$ such that $q' \in Q'$. We write $\delta(q)$ for the set of successors of q; thus,

$$\delta(q) = \bigcap_{a \in \Sigma} \bigcup_{Q \in \delta(q,a)} Q$$

A computation of S is an infinite sequence $\lambda = q_0, q_1, q_2 \ldots$ of states such that (for each i) q_{i+1} is a successor of q_i. We write $\lambda[0, i]$ for the finite prefix $q_0, q_1, q_2, \ldots, q_i$.

Often, we are interested in the cooperation of a subset $A \subseteq \Sigma$ of agents. Given A, we define $\delta(q, A) = \{\bigcap_{a \in A} Q_a \mid Q_a \in \delta(q, a)\}$. Intuitively, when the system is in state q, the agents in A can choose a set $T \in \delta(q, A)$ such that, no matter what the other agents do, the next state of the system is in T. Note that $\delta(q, \{a\})$ is just $\delta(q, a)$, and $\delta(q, \Sigma)$ is the set of singleton successors of q.

Example 1 ([4]). Consider a system with two agents "user" u and "telephone exchange" e. The user may lift the handset, represented as assigning value true to the boolean variable "offhook". The exchange may then send a tone, represented by assigning value true to the boolean variable "tone". Initially, both variables are false. Clearly, obtaining a tone requires collaboration of both agents.

We model this as an ATS $S = (Q, \pi, \delta, I)$ over the agents $\Sigma = \{u, e\}$ and propositions $P = \{\text{offhook,tone}\}$. Let $Q = \{00, 01, 10, 11\}$. 00 is the state in which both are false, 01 the state in which "offhook" is false and "tone" is true, etc. (thus, $\pi(00) = \emptyset$, $\pi(01) = \{tone\}$, etc.). The transition function δ and initial states I are as indicated in the figure.

$\delta(q, a)$	u	e
00	$\{\{00, 01\}, \{10, 11\}\}$	$\{\{00, 10\}\}$
10	$\{\{10, 11\}\}$	$\{\{00, 10\}, \{01, 11\}\}$
01	$\{\{00, 01\}, \{10, 11\}\}$	$\{\{01, 11\}\}$
11	$\{\{10, 11\}\}$	$\{\{01, 11\}\}$
I	$\{\{00, 01\}\}$	$\{\{00, 10\}\}$

Fig. 1. The transition function of the ATS.

2.3   Semantics

The semantics of ATL uses the notion of *strategy*. A *strategy* for an agent $a \in \Sigma$ is a mapping $f_a : Q^+ \to 2^Q$ such that $f_a(\lambda \cdot q) \in \delta(q, a)$ with $\lambda \in Q^*$. In other words, the strategy is a recipe for a to make its choices. Given a state q, a set A of agents, and a family $F_A = \{f_a \mid a \in A\}$ of strategies, the *outcomes* of F_A from q are the set $out(q, F_A)$ of all computations from q where agents in A follow their strategies, that is,

$$out(q_0, F_A) = \{\lambda = q_0, q_1, q_2, \cdots \mid \forall i, \ q_{i+1} \in \delta(q_i) \cap (\bigcap_{a \in A} f_a(\lambda[0, i]))\}.$$

If $A = \emptyset$, then $out(q, F_A)$ is the set of all computations, while if $A = \Sigma$ then it consists of precisely one computation.

The semantics of ATL* is as CTL*, with the addition of:

- $q \vDash \langle\!\langle A \rangle\!\rangle \psi$ if there exists a set F_A of strategies, one for each agent in A, such that for all computations $\lambda \in out(q, F_A)$ we have $\lambda \vDash \psi$.

Remark 1. To help understand the ideas of ATL, we state below some validities, and more surprising non-validities.

1. If $A \subseteq B$, then $\langle\!\langle A \rangle\!\rangle \psi \to \langle\!\langle B \rangle\!\rangle \psi$, and $[[B]]\psi \to [[A]]\psi$. Intuitively, anything that A can enforce can also be enforced by a superset B; and if anything that B is powerless to prevent cannot be prevented by a subset of B.

2. In CTL, A distributes over $\wedge$. But in general, $\langle\!\langle A \rangle\!\rangle (\psi_1 \wedge \psi_2)$ only implies $(\langle\!\langle A \rangle\!\rangle \psi_1) \wedge (\langle\!\langle A \rangle\!\rangle \psi_2)$. The first formula asserts that A can enforce $\psi_1 \wedge \psi_2$, while the second is weaker, asserting that A has a way to enforce ψ_1 and another, possibly incompatible, way to enforce ψ_2. Similarly, $\langle\!\langle A \rangle\!\rangle (\psi_1 \vee \psi_2)$ and $\langle\!\langle A \rangle\!\rangle \psi_1 \vee \langle\!\langle A \rangle\!\rangle \psi_2$ are different (for $A \neq \Sigma$). The first one asserts that A can enforce $\psi_1 \vee \psi_2$, but which of the two is true might be chosen by others. This is weaker than the second formula, which asserts that A can guarantee ψ_1, or A can guarantee ψ_2, but nobody can choose which. The strongest variant where A can choose, is expressed as: $\langle\!\langle A \rangle\!\rangle (\psi_1 \wedge \neg\psi_2) \wedge \langle\!\langle A \rangle\!\rangle (\psi_2 \wedge \neg\psi_1)$.

3. By repeating a cooperation inside a temporal operator, we weaken the formula, for instance: $\langle\!\langle A \rangle\!\rangle \Box \Diamond \phi \to \langle\!\langle A \rangle\!\rangle \Box \langle\!\langle A \rangle\!\rangle \Diamond \phi$. This is because the strategies F_A that A use in the outer modality may be adapted for the inner modality, by shifting its time: each $f'_a(x)$ is simply $f_a(\lambda \cdot x)$, where λ is the path linking the points of evaluation of the two modalities. (Note the CTL* validities $\mathsf{E}\Box\Diamond\phi \to \mathsf{E}\Box\mathsf{E}\Diamond\phi$ and $\mathsf{A}\Box\Diamond\phi \leftrightarrow \mathsf{A}\Box\mathsf{A}\Diamond\phi$.)

2.4   Guarded command language

ATSs may be described using a Mocha-like guarded command language. (Mocha [1] is the system modelling language used for ATL.) We illustrate this with the system S of the preceding section.

```
agent USER
  controls offhook;
  init
     offhook := false;
  update
     true -> ;
     true -> offhook := true;
endagent;
```

```
agent EXCH
  controls tone;
  init
     tone := false ;
  update
     true -> ;
     offhook -> tone := true;
endagent;
```

The init clause gives the initial values of variables (if they are not mentioned, their initial values are selected non-deterministically). The update clause consists of a set of guarded commands, consisting of a guard (before the arrow) and a command (after the arrow). The agents are run in parallel. At each step, the guards in the agent are evaluated, and the agent chooses one which evaluates to true. The command corresponding to that guard is executed. If a variable is not assigned to in a command, it preserves its old value. In particular, if the command is empty, nothing changes: the cryptic-looking command true -> simply allows the user to wait. Every variable is controlled by precisely one agent; only the controlling agent can assign to the variable. Agents may refer to variables which are controlled by other agents (for example, EXCH refers to offhook which is controlled by USER).

2.5   Simulation and trace containment

It is known in CTL that if a transition system S' *simulates* another one S, written $S \leq S'$, then all ACTL* formulas which hold of S also hold of S'. (ACTL* is the universal fragment of CTL*, i.e. the fragment in which the only path quantifier is A, and no negations are allowed which include A in their scope.)

A similar result holds for ATL* [3]. Instead of a single notion of simulation, they define a notion indexed by a set of agents A. Let $S = (Q, \pi, \delta, I)$ and $S' = (Q', \pi', \delta', I')$ be ATSs over agents Σ, with $P \subseteq P'$. For a subset $A \subseteq \Sigma$ of agents, a relation $H \subseteq Q \times Q'$ is an A-*simulation* from S to S' if[2]:

- For every set $T \in I(A)$, there exists a set $T' \in I'(A)$ such that for every set $R' \in I'(\Sigma - A)$ there exists a set $R \in I(\Sigma - A)$ such that $(T \cap R) \times (T' \cap R') \subseteq H$.

and, for all states q, q' with $H(q, q')$, we have

- $\pi(q) = \pi'(q') \cap P$;
- For every set $T \in \delta(q, A)$, there exists a set $T' \in \delta'(q', A)$ such that for every set $R' \in \delta'(q', \Sigma - A)$ there exists a set $R \in \delta(q, \Sigma - A)$ such that $(T \cap R) \times (T' \cap R') \subseteq H$.

---

[2] Our definition slightly generalises that of [3] by allowing multiple initial states and new propositions and agents.

The intuition is that whatever A can do in S, A can also do it in S' so that whatever the other agents do in S', they could already do it in S to yield a similar state. Intuitively, S' conserves all the capabilities A has in S, perhaps adding some more.

We say that S' A-simulates S, and write $S \leq_A S'$, if there is a simulation from S to S'. Intuitively, this holds if A has a superset in S' of the capabilities it has in S. It is proved in [3] that $S \leq_A S'$ iff every $\langle\!\langle A \rangle\!\rangle$-ATL* formula satisfied by S is also satisfied by S'. This formalises the intuition just mentioned, since formulas in $\langle\!\langle A \rangle\!\rangle$-ATL* assert capabilities of A.

3   Features and the feature construct

Our goal in this paper is to show how certain properties can be preserved through the addition of features. From this, we can demonstrate feature non-interaction, as explained in the introduction.

Our approach is to define a *feature construct* for the Mocha-like guarded command language introduced in section 2.4. The feature construct plays a similar role to the one defined for SMV [18]; it is also similar to the idea of superimposition [12]. Using it, we give examples of features and show, for specific features, that the system without the feature is an A-simulation of the system with the feature. From this, we conclude that properties of the base system are inherited by the system with features.

This section is structured as follows. In section 3.1 we model a Plain Old Telephone System (POTS) and some of its properties. Section 3.2 defines the feature construct, and gives some examples for POTS. We then study feature interactions in section 3.4.

3.1   POTS and its properties

Example 2. A more complete POTS model is defined using the guarded command language of section 2.4. In figure 2, we model the user: she may cause the phone to go offhook or onhook at will (**nondet** is a shorthand for a choice among all possible values of the type), and while the phone is offhook she may dial a number.

In figure 3, we model the exchange (without technical details). It consists of n identical agents, one for each user. It has a variable st, for status, which is initially idle. When the user goes offhook, st becomes dialt, for dialtone. If st is idle and another person tries to ring us, st becomes ringing, and we note the identity of the caller. If two users i, j simultaneously ring a third one k, the exchange must arbitrate by choosing one of them to succeed (gets ringing-tone) and the other one to fail (gets busy-tone). The exchange does this by setting ex[k].caller to i or to j.

The system consists of an array of exchanges and an array of users. Notice the parameter for EXCH: it is given the value of its own number, which it calls s (for 'self').

```
agent USER
controls
   offhook : boolean;
   dialed : Number;
init
   offhook := false;
update
   offhook -> dialed := nondet;
   -> offhook := nondet;
endagent;
```

Fig. 2. Code for USER

```
agent EXCH (s)
controls
   st : {idle, dialt, trying, busyt, ringingt, talking,
                              ringing, talked, ended };
   callee : Number;
   caller : Number;
init
   st' := idle;
update
   user[s].offhook & !user[s].offhook' -> st'=idle;
   st=idle & user[s].offhook' -> st' := dialt;
   st=idle & ex[j].callee=s &
                 ex[j].st=trying & !user[s].offhook'
                 -> st' := ringing; caller' := j;
   st=dialt & user[s].offhook'
                 & user[s].dialed'=n -> callee' := n;
   ⋮
   st=trying & callee=j & ex[j].st=idle & ex[j].caller'=s
                       & user[s].offhook' -> st' := ringingt;
   st=trying & callee=j & ex[j].st=idle & ex[j].caller'!=s
                       & user[s].offhook' -> st' := busyt;
   st=trying & callee=j & !ex[j].st=idle
                       & user[s].offhook'-> st' := busyt;
   ⋮
endagent
```

Fig. 3. Code for EXCH

```
ex : array 1..n of EXCH;
ex[i] := EXCH(i);
user[i] := USER
```

Fig. 4. Code for POTS

The logic ATL is well-suited for expressing specifications of telephone systems, because the users are autonomous, and we are interested in whether they have the power to enforce certain behaviours. Compared with the properties defined using CTL in [18], ATL offers us the opportunity to distinguish between different sources of non-determinism, which makes the specification reflect our intentions more precisely. We illustrate with a few examples:

1. *Any phone may call any other phone.* In [18] this was approximated in CTL:
   $$\forall i \neq j.\ A\square E\lozenge(\text{ex}[i].\text{st=talking \& ex}[i].\text{callee=}j)$$
   indicating that, in all reachable states, there is a path which eventually leads to i and j talking to each other. This is rather weaker than the intention, which was that it is within i's and j's joint power that i initiate a successful call to j. We may express that as $\forall i \neq j$
   $$A\square\langle\langle\text{user}[i],\text{user}[j]\rangle\rangle\lozenge(\text{ex}[i].\text{st=talking \& ex}[i].\text{callee=}j)$$
   A similar formula which is slightly weaker but has the advantage of being within $\langle\langle\text{user}[i],\text{user}[j]\rangle\rangle$-ATL is $\forall i \neq j$:
   $$\langle\langle\text{user}[i],\text{user}[j]\rangle\rangle\square\lozenge(\text{ex}[i].\text{st=talking \& ex}[i].\text{callee=}j)$$

2. *The user cannot change the callee without replacing the hand-set.* In [18] it is expressed in CTL as:
   ```
   A□ ((ex[i].callee=j & ex[i].st=trying)
            -> (A[ ex[i].callee=j W ex[i].st=idle ]))
   ```
   This is rather stronger than the intention: this forbids any change of callee. This CTL formula becomes false in the context of call-forwarding, where the system may change the callee as i sets up the call. In ATL, we capture the requirement more precisely:
   ```
   A□(ex[i].callee=j & ex[i].st=trying
            -> [[user[i]]](ex[i].callee=j W ex[i].st=idle))
   ```
   This weaker formula is true even if the system can change the callee. Again, a slightly weaker formula in [[user[i]]]-ATL is possible:
   ```
   [[user[i]]]□(ex[i].callee=j & ex[i].st=trying
            -> (ex[i].callee=j W ex[i].st=idle))
   ```

3.2   Feature construct definition

The feature construct that we use here is an adaptation of the generic idea of [18]. The base language that we use is a simplification of the *Reactive Modules* formalism [2] used by Mocha [1], that we presented in section 2.4.

Following [18], a feature can be seen as a prescription for changing a basic system. That which is assumed of a basic system will appear in the **require** section of the feature. Here, we can require particular agents and variables. The feature will add to the system new variables and agents to deal with the feature in the **introduce** section. Because many features need to be activated before taking effect, we usually introduce a boolean variable **use** that indicates whether the feature is activated. Finally, the **change** section indicates how the behaviour of the existing system is changed. Currently, we have four types of changes:

1. **if** *condition* **then override** c means that when *condition* is evaluated to true, the existing commands are disabled, and only the command c is allowed to execute.

2. **if** *condition* **then expand** c means that when *condition* is evaluated to true, the command c is allowed to execute. The existing commands are still enabled as before: the non-determinism of the system is increased.

3. **if** *condition* **then impose** c means that when *condition* is evaluated to true, the command c, which is a set of parallel assignments $x' := e$, determines the new values of these x variables. The values of other variables are set by an existing command.

4. **if** *condition* **then treat** c means that when *condition* is evaluated to true, the command c, which is a set of parallel assignments $x := e$, is used to determine the value of x in expressions. The variable x still exists, and will be accessible again when the condition reverts to false.

Only the last two types were present in [18]. The first two types can also be defined both in terms of syntactic manipulations or semantically, on the agent's transitions.

Finally a feature comes with **properties**, that describe its essential functionalities in a high-level way. These properties need not exhaustively specify the system. The specifier is intended to write properties which should be preserved when this feature is combined with other features. In this paper, we advocate the use of ATL* for properties.

Example 3. It is now very common to have many features on top of POTS. These features come in many variants, and are now being standardised [8].

For instance, the feature Call Forward When Busy (CFB) adds the following typical behaviour: When CFB is active and the subscriber's line is busy, incoming calls are diverted to a phone number pre-specified by the subscriber. The number can be changed, and the feature can be enabled or disabled at subscriber's will. The feature is implemented by changing the exchange of the caller, and adding new commands to the subscriber i, see fig. 5.

The fundamental property of forwarding is that user j can ensure that any user who tries to reach him will try user k instead, and j can choose any k. Note the scope of the quantifications (cf. remark 1.2).

Example 4. The feature Ring Back When Free (RBWF) also avoid the annoyance of busy callees, but this time it is a feature of the caller (me, say): If I get the busy tone when calling a number, I can activate RBWF. RBWF will then attempt to establish a connection as soon as the callee is free. It first calls me with a special ring; when I then lift the handset, a call is initiated on my behalf.

To model this (see fig. 6), we introduce **awaited**, the number we are trying to reach. Since we introduce a single number, only the last RBWF may be pending. Also we use Mocha's notion of **event** to model activation: it is

```
feature CFB(i)
require ...
introduce
  agent USER[i]
    controls
        use  : boolean
        forw : Number
    init
        use := false
change
  agent USER[i]
    expand use  := nondet;
    expand forw := nondet;

  agent EXCH(i)
    if st = trying & callee = i  & user[i].use & ex[i].st != idle
    then override callee' := user[i].forw;

properties
```

$\forall k. \langle\!\langle \text{user}[i] \rangle\!\rangle \Diamond \Box \forall j. (\text{ex}[j].\text{st=trying} \, \& \, \text{ex}[j].\text{callee=}i \, \& \, \text{ex}[i].\text{st!=idle}$
$\rightarrow \text{ex}[j].\text{st=trying U (ex}[i].\text{st=trying} \, \& \, \text{ex}[i].\text{callee=}k))$

Fig. 5. Call Forward when Busy

an instantaneous action, whose occurrence can be caused by **event!** (equivalent to toggling **event**) and tested by **event?** (equivalent to the condition **event=event'**).

Let us have a closer look at the properties: the first one simply says that users together can make my ringback scenario succeed: I hear the special ringing, then I take the phone offhook and call j. The collaboration of all users is needed for this success:

- i must of course enable the feature.
- j must agree to be first busy, then idle.
- The collaborations of other users is needed as well, since they could conspire to hold i or j busy all the time.

The user i alone is much less powerful: He might decide not to use the feature at all, by not setting **activate**. (Indeed, the fact that the user can avoid using the feature is important to our main result, section 3.4.)

This leads to a natural categorisation of features, similar in motivation to [7], but different in detail: features can be categorised according to the set of players that occur in the cooperation modality of the ATL formula of their properties. This essentially says who is in control of the feature. Specifically, we can distinguish single-user features, two-users features, group features, system features (where system is a specific player).

```
feature RBWF(i)
require ...
introduce
 agent USER(i)
   event activate
 agent EXCH(i)
    controls
        use : boolean
        awaited : Number
        special_ring : boolean
     init
       use := false;
       special_ring := false;

change
 agent USER(i)
   expand activate!
 agent EXCH(i)
 if st = busyt & user[i].activate?
 then impose use' := true ; awaited' := dialed;

 if use & st = idle & ex[awaited].st = idle
 then override callee' := awaited;
                 st' := ringing;
                 special_ring' := true;

 if use & st = ringing & special_ring & user[i].offhook'
 then override st' := trying;
                 special_ring' := false; use' := false
```

properties
   $\langle\langle\text{user}\rangle\rangle \Diamond((\text{ex[i].st = ringing \& ex[i].special\_ring})$
          $U~(\text{user[i].offhook U ex[i].talking \& ex[i].callee = j}))$
   $\langle\langle\text{user}[i]\rangle\rangle~!\text{ex[i].use}$
   $\langle\langle\text{user}[i]\rangle\rangle\Diamond((\text{ex[i].use \& ex[i].awaited=j}) \mid (\text{ex[i].st=ringingt}))$

Fig. 6. Ring Back When Free

3.3   Feature Construct Semantics

We define the semantics of the feature constructs **override, impose, expand** and **treat** by syntactic transformation of the Mocha-like language. Dealing with the **require** and **introduce** sections is straightforward: for **require**, we check that the required items are present (the feature integration fails if they are not), and for **introduce** we simply add the new data.

The **change** section is dealt with as follows. Suppose we start with the program in figure 7, and we integrate a feature.

```
agent A
controls
  ...
init
  ...
update
  g1 -> c1;
  g2 -> c2;
   :
  gn -> cn;
endagent
```

Fig. 7. Some arbitrary code for an agent A.

- For the feature **if** g **then impose** $x := e$, the update section of the program becomes:

```
g1 & !g -> c1;
g1 & g -> c1 [x:=e];
g2 & !g -> c2;
g2 & g -> c2 [x:=e];
   :
gn & !g -> cn;
gn & g -> cn [x:=e];
```

The meaning of $c[x := e]$ where c is a set of assigments is to replace (if present) the assigment of x in c by the new one $x := e$, or to add it (if not present). (Recall that in Mocha the list of assignments are performed simultaneously.)

- For the feature **if** g **then override** c, the update section of the program becomes:

```
g1 & !g -> c1;
g2 & !g -> c2;
   :
gn & !g -> cn;
g -> c;
```

- For the feature **if** g **then expand** c, the update section of the program becomes:

```
g1 -> c1;
g2 -> c2;
    ⋮
gn -> cn;
g -> c;
```

- For the feature **if** g **then treat** $x = f$, the update section of the program becomes:

```
g1 -> c1';
g2 -> c2';
    ⋮
gn -> cn';
```

where x_i' is c_i but with x replaced with the conditional expression $g?f : x$ (i.e. if g then f else x).

3.4 Feature interactions

Thanks to the properties that are part of our features, we can define interactions as a discrepancy between the expected properties of the system with features and the actual ones. We note a feature as (F, ϕ) where ϕ is the properties sections and F is the description of how the feature is implemented. Applying the feature F to a system S satisfying its requirements will be denoted $S + F$. This operation is also called "feature integration". We assume that the **require** section, the **change** section, and the introduced properties are consistent with each other: that is, that $S + F \models \phi$ for any S satisfying the requirements.

Now we can define a feature interaction as non-preservation of the properties of integrated features:

- The feature F interacts with the system S by destroying some core property ϕ_S of the system: $S + F \not\models \phi_S$;
- The feature F_2 interacts with the feature F_1 by destroying a property ϕ_1 introduced by F_1: $S + F_1 + F_2 \not\models \phi_1$.

The goal of feature-oriented programming is to be able to produce rapidly systems with a large number of features integrated, and to ensure the absence of feature interactions for such systems.

It is thus important to prove generic preservation properties: a feature F preserves all properties of a class C if (for all $\phi \in C$) $S \models \phi$ implies $S + F \models \phi$.

We have seen that simulation relations are the right tool to this end: they ensure that a wide class of properties are preserved when adding a feature. These relations give a precise meaning to the notion of backward compatibility.

In particular, if we can show that $S + F$ A-simulates S (for any S), when integrating F in a new system, we know that many properties of this new system do not need to be checked. Features are usually intended to augment the power of their users: formulas talking about these powers are thus preserved.

Proving this property of F can sometimes be done easily. First, we define an *A-enabled variable* **use** *introduced by* F (where $A \subseteq \Sigma$) to have the following properties:

- the variable is **introduced** by F in some agent a
- the variable is initially false: **use:=false** appears in an **initintroduced** by F.
- the variable can only be set by agents in A using **expanded** commands. This can often be checked syntactically: For instance, if **use** is only set to true by an **expanded** command of an agent in A, as in CFB, this is immediate. In RBWF, this is indirect: **use** is controlled by **exch[i]** (an agent outside A) but the guard contains a variable **activate** controlled by an **expanded** command of user[i]. More generally, this can be verified by checking that the feature F', which is F without the **expand** of agents in A, satisfies A□¬**use**, by which we mean that for any base S, $S + F' \models$ A□¬**use**.

This condition can be used to ensure that, if agents in A behave exactly as they did in the old system, they will not enable the feature.

Theorem 1. If all changes of F are of one of the following forms:

- a change that is guarded by an A-enabled variable introduced by F
- an **impose** where all affected variables are **introduced** by F.
- an **expand** of an agent in A.
- "**if** $g \wedge g'$ **then override** c" in an agent in $\Sigma - A$, if $g \rightarrow c$ is a command of this agent.

then F is A-preserving, i.e. $S + F$ A-simulates S for any S that satisfies the **require** clause.

The idea of the proof is to note that the relation obtained by requiring that all old variables have the same value, and that the enabling variables **use** are false, is an A-simulation from S to $S + F$. Thus all properties written in $\langle\!\langle A \rangle\!\rangle$-ATL are preserved. We cannot give a real proof of this theorem here, as it requires the precise semantics of our Mocha-like language which is not given in this paper.

Example 5. The features CFB(i) and RBWF(i) are A-preserving for any set of agents A containing user[i]. Since their properties are also in this fragment, these features will not interact (in the sense of this paper).

Note that it is usually considered that these features do interact, since a user A that called B, was forwarded to C and activated RBWF might well end up calling back C in some implementations, while he probably intended to call B. Here this interaction is correctly, but silently, handled by our model, since it does not belong to the class of interactions defined in this paper.

3.5   Preliminary Experiments

Currently, there is no automatic translation from our Mocha-like language to Mocha. However, we have successfully implemented the model of the POTS given in Fig. 2 to 4 of section 3.1 with 4 users, and checked the properties discussed in section 3.1.

4   Conclusions

We have shown a general case in which introducing a feature provably does not break a class of properties: this holds when integrating the feature results in a U-simulation of the original system for some group of users U, and the properties assert capabilities of the users in U. We have indicated four types of changes that are U-preserving. We illustrated with examples from the telephone system. Most telephone features naturally fit into one of the cases of the theorem. Thus the proofs of non-interaction that [18] had to perform for all combination of features can now be obtained by a simple, single syntactic check.

The general technique, in principle, can work for any logic and its associated notion of simulation. However, we have found that ATL* provides a rich set of fragments and associated simulations, that are suited to the application domain: features are valuable only because they offer new capabilities to their users, and thus their properties are naturally expressed in ATL*. Actually, our example properties were all in the smaller fragment $\langle\!\langle A \rangle\!\rangle$-LTL[3], for which the weaker $\langle\!\langle A \rangle\!\rangle$-trace containment suffices. We didn't pursue this line of research since all our features happen to be preserving also the stronger $\langle\!\langle A \rangle\!\rangle$-simulation, and this preservation is easier to show.

The special case where $U = \emptyset$ allows to show the preservation of invariants of the system (or more generally, ACTL* formulas). However, the corresponding simulation only allows features to make their agents more deterministic, which is rarely useful.

We have seen intuitively appealing properties of the form $A\square\langle\!\langle U \rangle\!\rangle \Diamond \phi$. Our method could be extended by discovering the "simulation" relation corresponding to these formulas, and looking for a simple way to prove that a feature preserves this relation. We plan to define a suitable notion of U-resettable systems. Intuitively, the telephone system is resettable by its users: if they all hang up and switch off their features, the system returns to its initial state. We would like to define this precisely, and prove of U-resettable systems that $\langle\!\langle U \rangle\!\rangle \Diamond \phi$ is equivalent to $A\square\langle\!\langle U \rangle\!\rangle \Diamond \phi$ (from left to right is done by prefixing the strategy with a reset). This would imply that formulas of the form $A\square\langle\langle U\rangle\rangle\Diamond\phi$ are preserved by U-conservative features.

The idea of this paper may be seen as a special case of a proof rule of the form

$$\frac{S \vDash \phi \quad \text{condition on } F, \phi}{S + F \vDash \phi}$$

which allows us to preserve the property ϕ through the addition of the feature F. In this paper, the condition on F, ϕ is that F is U-conservative and ϕ is in $\langle\langle U\rangle\rangle$-ATL*. Other conditions on F, ϕ can be used. In another paper, we are modelling features as warps in the transition system and deriving from ϕ a simpler formula which the warp is required to preserve [11].

A related problem is to show the internal consistency of features, by which we mean that $S + F \vDash \phi$ for any S that satisfies the requires clause. By inserting the needed properties in the requires clause, the combination of features could eventually become a matter of plug and play, with well defined and easily combinable compatibility properties.

Finally, we used here only two levels for describing a property: the level of models, and of formulas. Lower levels indicating how to integrate features at the level of code would make the approach practical, and checking consistency between levels will improve our confidence in features.

Achkowledgments. The three authors are members of the FIREworks[3] Esprit Working Group, and gratefully acknowledge support for travel which enabled them to meet together. Mark Dermot Ryan also acknowledges British Telecom for generous support, and Pierre-Yves Schobbens thanks the University of Birmingham for funding an invited professorship that provided a further opportunity to work on this material.

References

1. R. Alur, H. Anand, R. Grosu, F. Ivancic, M. Kang, M. McDougall, B.-Y. Wang, L. de Alfaro, T. Henzinger, B. Horowitz, R. Majumdar, F. Mang, C. Meyer, M. Minea, S. Qadeer, S. Rajamani, and J.-F. Raskin. *Mocha User Manual.* University of California, Berkeley. www.eecs.berkeley.edu/~mocha.
2. R. Alur and T. Henzinger. Reactive modules. *Formal Methods in System Design,* 15(1):7–48, 1999.
3. R. Alur, T. Henzinger, O. Kupferman, and M. Vardi. Alternating refinement relations. In D. Sangiorgi and R. de Simone, editors, *CONCUR 98: Concurrency Theory,* Lecture Notes in Computer Science 1466, pages 163–178. Springer-Verlag, 1998.
4. R. Alur, T. A. Henzinger, and O. Kupferman. Alternating-time temporal logic. In *Proceedings of the 38th Annual Symposium on Foundations of Computer Science,* pages 100–109. IEEE Computer Society Press, 1997.

---

[3] www.cs.bham.ac.uk/~mcp/fireworks/

5. M. Calder and E. Magill, editors. *Feature Interactions in Telecommunications and Software Systems VI.* IOS Press, 2000.

6. M. Calder and S. Reiff. Modelling legacy telecommunications switching systems for interaction analysis. In *Systems Engineering for Business Process Change.* Springer Verlag.

7. E. Cameron, N. Griffeth, Y.-J. Lin, M. Nilson, W. Schnure, and H. Velthuijsen. A feature interaction benchmark for in and beyond. In W. Bouma and H. Velthuijsen, editors, *Feature Interactions in Telecommunication Systems.* IOS Press, 1994.

8. CCITT. *Recommendation Q.1215, Distributed Functional Plane for Intelligent Network CS1.*, 1992.

9. E. M. Clarke and E. A. Emerson. Synthesis of synchronization skeletons for branching time temporal logic. In D. Kozen, editor, *Logic of Programs Workshop*, number 131 in LNCS. Springer Verlag, 1981.

10. E. M. Clarke, O. Grumberg, and D. A. Peled. *Model Checking.* MIT Press, 1999.

11. H.-D. Ehrich, M. D. Ryan, and P.-Y. Schobbens. Preserving temporal properties through time warps. In preparation.

12. S. Katz. A superimposition control construct for distributed systems. *ACM Transactions on Programming Languages and Systems*, 15(2):337–356, April 1993.

13. K. Kimbler and L. G. Bouma, editors. *Feature Interactions in Telecommunications and Software Systems V.* IOS Press, Sept. 1998.

14. M. Kolberg, E. Magill, D. Marples, and S. Reiff. Results of the second feature interaction contest. In Calder and Magill [5], pages 311–325.

15. H. Korver. Detecting feature interactions with CÆSAR/ALDÉBARAN. *Science of Computer Programming*, 29(1–2):259–278, July 1997.

16. K. L. McMillan. *Symbolic Model Checking.* Kluwer Academic Publishers, 1993.

17. M. Plath and M. D. Ryan. Entry for FIW'00 Feature Interaction Contest. Technical report, School of Computer Science, University of Birmingham, 2000. Available from www.cs.bham.ac.uk/~mdr/papers.html. Also summarised in [14].

18. M. C. Plath and M. D. Ryan. Feature integration using a feature construct. *Science of Computer Programming*, 2000. To appear. A shorter and earlier version of this paper appeared in [13], pages 150–164.

Algebraic Treatment of Feature-oriented Systems

Christophe Gaston, Marc Aiguier, and Pascale Le Gall

L.a.M.I., Université d'Évry, Cours Monseigneur Roméro, 91025 Évry, France
{gaston,aiguier,legall}@lami.univ-evry.fr

Abstract. An important aspect of the feature interaction problem is to formally capture the notion of feature interactions. Although this notion is quite well informally understood by the researchers of the domain, the way, they handle it, strongly depends on the field of investigation they decide to work on (formal method application, architectural conception, technological research...). In this article, we focus on how formally specifying and studying feature systems, and both integration and interaction of features. More precisely, we aim to give a logic-independent framework to deal with the notions of feature, feature-based systems and feature interactions. Then, to help the reader's intuition, we instantiate it by a dynamic algebraic formalism and we give concrete examples of interactions between two features previously described in this formalism.

Key words : feature, feature interaction, formal specification, abstract logical framework, dynamic logic, algebraic techniques.

1 Introduction

The notion of features has emerged in the telecommunication systems field [6,7,14]. Software telecommunication systems are often composed of a "kernel" providing the basic expected functionalities and a set of satellite entities, called *features*. Each of these aims to modify the set of functionalities characterising the rest of the system (possibly including other already existing features). In fact, features aim to "enrich" the possibilities of the users : any user can easily enrich his own set of possibilities by subscribing to his personalised package of features.

From the design point of view, it is clear that such systems have to allow an easy updating in order to provide the capacities promised by any upcoming feature : This updating is classically called "feature integration".
Features are intended to be separately designed, implemented and integrated. Actually, a feature is often described on a subpart of the system description. This subpart denotes a set of requirements which are those dealing with the behavior of the system that the feature of interest aims to update. More precisely, a feature may either, add or modify some functionalities, or combine those two aspects. Thus, the description of a feature does not only consist of the specification of the functionalities it offers and/or modifies but also contains some required properties about the system it is supposed to be integrated on later. Practically, among these last properties, there is particular

emphasis on the basic properties that the feature ensures to preserve when integrated on the rest of the system. They represent the essence of the underlying system to be completed by the integration of some features. Thus, they may be considered as some invariants of the system under consideration. For example, in software telecommunication systems, the fact that each portion of a communication established between users should be paid at least by one of them, is invariably shared by most of the well-known features.

In general, a feature is easily understood, designed and validated in isolation. Difficulties arise when several features are integrated together on some basic system: it induces mutual influences which find expression in emergence or loss of some behaviors of the system. Let us point out that, according to the cases, those feature interactions are desirable or not. The decision concerning the qualitative nature of the detected interactions is an expert matter. Due to the intrinsic flexibility of feature-oriented systems, the integration of a new feature may lead to modify the previously specified system. Thus, when an undesirable interaction is detected, it re-opens the previous design decisions.

Formal methods seem promising in helping to bring solutions to the feature interaction problem [8]. They have already been intensively used, in particular through model-checking techniques [16]. Let us cite some of their advantages: they are useful in giving strong hints about the nature of interactions. Indeed, they give some theoretical foundations to classify such interactions. Moreover, they allow to take advantage of some associated tools as automatic analysis, rapid prototyping, simulation or test case generation.

The field of feature-oriented systems is now mature enough to consider the possibility of formally defining feature-oriented systems in a generic way. The main contribution with such an approach is to provide for both designers and users of logics, a methodology giving a way for specifying feature systems and studying their interactions systematically (when the logic is appropriated for this purpose. See Definition 2 for more explanations). Thus we prevent the authors and users of such logics from developing lots of unavoidable formal definitions and results (as defining feature, feature integration and feature interactions) in their formalism. Moreover, assuming that a feature-oriented system is simply any system characterised by the set of services it offers to its users, from our point of view, features can be seen as a new paradigm for designing systems, not only the software telecommunication systems but also any kind of software systems. In this way, we can make an analogy between feature-oriented systems and both object-oriented and modular hierarchical systems. Besides, for each of them, logic-independent frameworks have already been introduced to generically define the key notions of the corresponding design style. Regarding to modular systems, general concepts of modularity [9,15,4,5] have often been defined using institutions [10] as underlying meta-formalism. Since the main characteristics of modular systems is the preservation of the imported elements (e.g. models or properties), institution-like frameworks seem really adequate to define general modular systems because the satisfaction condition ensures a preservation principle through signature morphisms that modelise interface inclusions, renaming or hiding. Regarding to objects, general concepts [17] have been defined using a particular class of institutions, more or less extensions of the temporal logic. Indeed, as an object is essentially given by an identity and a state, [17] specialises the logical background in order to fit with their will of modelling objects. More simply, thanks to a high level of abstraction, all those approaches allow to deeply understand the underlying principal concepts set in action. Those works introduce guidelines which can be easily instantiated in any concrete specification framework related to the concerned paradigm.

In this paper, our purpose is then to propose a first logic-independent framework dedicated to modelise the notion of *"feature-oriented systems"*. The use of an algebraic setting allows us to formally characterise features, feature integration and feature interactions. In order to define such a generic feature-oriented framework, we need an adequate underlying meta-formalism in the spirit of modular and object-oriented approaches mentioned above. For this, we modify the very classical institutions framework, in accordance with the needs issued from the modelling of features. As usual, our formalisms will consist of a syntactic part, the signatures and the sentences which represent respectively interface systems and properties about them, and for each signature, a semantic part which consists of both a class of models and a satisfaction relation which respectively denote all the possible systems and validation between models and sentences. The main modifications are the following :

- Most of feature interactions emerge with a sharing of identifier symbols which is not well controlled. For example, when using the model-checker SMV for detecting feature interactions, [16] emphasises on a strict use of the involved variables: any variable useful to describe the considered feature should be declared either as a new one or as already present in the basic system. In order to express this potential sharing of identifiers between features, we provide signatures with the appropriate notions issued from the set theory as inclusion, intersection, union ... From a technical point of view, we simply reuse the strongly inclusive category introduced in [9] for modular systems.
- Another significant modification we make on institutions framework is that we do not impose the satisfaction condition. Indeed, contrary to the modular approach based on the preservation of properties inherited from imported sub-systems, feature-oriented approach is based on the possibility of revising the behavior of the underlying system by integrating features on it. Thus imposing such a condition may prevent us to detect some interactions which could be harmful.

The paper is organised as follows: Section 2 introduces basic requirements we expect from a formalism to be used for feature paradigm. We define in a logic independent way the notions of *feature, feature system, feature integra-tion* and *feature interaction* in Section 3. In Section 4, we propose an example of formalism to illustrate our general framework. Section 5 is devoted to give examples of both constraint and emerging interactions from a specification previously described with the formalism of section 4. We leave recapitulation and some brief comments on future works to section 6.

2   Logics adapted for feature oriented conception

Herein, we start by defining the basic requirements that a formalism has to satisfy.

In order to define a generic version (i.e. logic-independent) of the feature paradigm, we follow basic ideas developed in the theory of institutions [10]. The theory of institutions abstracts the semantical part of the informal notion of logical systems. Intuitively, a logical system is given by a syntax, a semantic and a proof calculus. Syntax allows us to state properties denoting behaviors that we want to satisfy on a system. These properties have to be unmistakable in an unequivocal way. Then, they have to be defined formally. The whole set of properties to satisfy is usually called a *formal specification*, and the statement of a property is called a *sentence*. In practice, syntactic part of a logic is based on inductive constructions of formulas from symbols given in first. Classically, the set of these symbols is called a *signature*.

Symbolic treatment of properties is not sufficient to deal with the correctness of systems. Indeed, they are only unmeaning sequences of symbols. Then, we have to give a corresponding meaning "in the real world" to phenomena that we want to guarantee. To achieve this purpose, we have to rigorously modelise this real world, and then to consider mathematical models which represent its possible abstractions. Finally, we have to be able to denote among all sentences, properties satisfied by these models. These part is usually called *semantics*. This last part is fundamental in the sense that it allows to deal with correctness of systems scientifically. However, it is not sufficient. Indeed, a fundamental aspect of formal software engineering is to automatically check the correctness of softwares. So, the question is : How to automatically check (or prove) that a sequence of symbols is a property satisfied by our software. To reach this purpose, we have to define how to manipulate formulas and how these manipulations lead to obtain the truth of a property. We usually call these manipulations *inference rules*, and we talk about *calculus*.

Institutions only concern both syntactical and semantical aspects of a *log-ical system* To abstract the notion of *logical system*, [10] defines a signature as an element of a category. Then sentences are only elements of a set associated to a signature. All the computational contingencies and modelisations are not considered. In the same way, models associated to a signature are only ob-jects of a category. To characterise properties satisfied by a class of models, we define a relation, usually called *satisfaction relation* and denoted "$\models$", between models and sentences (see Definition 2 for the formal definition).

In order to design a software system as the set of features that it can offer, a logical framework devoted to this paradigm has to allow to put features together. However, a fundamental basic notion of the feature paradigm is that features are defined sharing some common informations issued from the basic system. Consequently, the joint integration of two features $\mathcal{F}_1$ and $\mathcal{F}_2$ has to take into account some sharing between them. At the signatures level, it means that the signature of a system including the two features $\mathcal{F}_1$ and $\mathcal{F}_2$ has to be constructed, taking into account the new elements introduced in the interface by those two features but also the elements (issued from the basic system) shared by the two features. Thus we need some kind of notion of union on signatures.

Classically, the categorical concept of pushouts and more generally finite colimits provides the abstract basis for combining signatures (and more generally specifications) in a systematic way [15].

It is well-known that the concept of pushouts is too general to deal with the concept of union, even though we take the category of sets usually denoted

Set. Indeed, given a pushout $\begin{array}{ccc} S_0 & \xrightarrow{g_1} & S_1 \\ \scriptstyle{g_2}\downarrow & & \downarrow \\ S_2 & \longrightarrow & S \end{array}$ in *Set*, arrows g_1 and g_2 may not

be inclusions (by the way nor injections) and then both S_0 may not be the intersection of S_1 and S_2 and S the union of them. Then, to deal with union, we have to parametrise the pushout by an object denoting the sharing between Σ_1 and Σ_2 (usually noted $\Sigma_1 \cap \Sigma_2$) provided with two morphisms denoting inclusions. However, the usual definition of union throughout a pushout is only up to isomorphisms. Intuitively, this means that this approach allows to rename symbols in the signature $\Sigma_1 \cup \Sigma_2$.

In our context, we prefer to avoid renaming. Indeed, a feature generally characterises the behaviour of a new functionality. Thus, its description introduces just a few new symbols. The greatest part of the symbols used refers to already existing functionalities, whose some behaviors have to be modified. To reach this result, we use the axiomatisation of the notion of inclusion given in [9]. The main advantage of this axiomatisation is that intersection and union are unique. Consequently, there is no more parametrisation by the intersection, and thus renaming is avoided (i.e. union is not defined up to isomorphism).

Let us recall the axiomatisation of the notion of inclusion given in [9] :

Definition 1. A category $\mathcal{C}$ is *strongly inclusive* if and only if $\mathcal{C}$ is a category with pullbacks and there exists two sub-categories $\mathcal{I}$ and $\mathcal{E}$ of $\mathcal{C}$ such that:

1. $|\mathcal{I}| = |\mathcal{E}| = |\mathcal{C}|$;
2. any $e \in Hom_{\mathcal{E}}$ is epic[1];

---

[1] A morphism is epic when it is right cancellable.

3. every morphism f in $\mathcal{C}$ can be factored uniquely as $\iota \circ e$ with $\iota \in \mathcal{I}$ and $e \in \mathcal{E}$;

4. $Hom_{\mathcal{I}}$ defines a partial order on $\mathcal{C}$ and the poset $(|\mathcal{C}|, Hom_{\mathcal{I}})$ is a lattice where the sup of A and B is the sum of A and B, denoted $A + B$, and the inf of A and B is the unique inclusion pullback in $\mathcal{C}$ of the sum, called *intersection* of A and B and denoted $A \cap B$ (both existence and uniqueness of $A \cap B$ are given in [9]). Moreover, we impose on $A + B$ to be the pushout of $A \cap B$.

The morphisms in $\mathcal{I}$ are called *inclusions* and the pair $(\mathcal{I}, \mathcal{E})$ is called the *inclusion system* of $\mathcal{C}$. Afterwards, we will use the notation $A \subseteq B$ to denote the morphism $\iota : A \to B$ in $Hom_{\mathcal{I}}(A, B)$ when this does not bring any ambiguity about.

Remark 1. Sum and intersection are unique and the full sub-category whose morphisms are inclusions is *finitely cocomplete* (see [9]).

As an example, the category *Set* provides with the inclusion system $(\mathcal{I}, \mathcal{E})$ where $\mathcal{I}$ contains all inclusions and $\mathcal{E}$ contains all surjections is triv-

ially strongly inclusive. Indeed, given a pushout

$$\begin{array}{ccc} S_0 & \xrightarrow{g_1} & S_1 \\ \scriptstyle g_2 \downarrow & & \downarrow \\ S_2 & \xrightarrow{} & S \end{array}$$

in *Set* with $S_0 = S_1 \cap S_2$ and all arrows inclusion, S denotes the set-theoretical union of S_1 and S_2 (i.e. $S = S_1 \cup S_2$). This pushout is also a pullback, so $S_1 \cup S_2$ is the pushout of $S_1 \cap S_2$.

Now let us define our *institution-like* framework to deal with feature oriented systems. Classically, an institution is represented by a category of signatures provided with, for each signature Σ, a set-valued functor yielding the Σ-formulas, a category-valued functor yielding the Σ-models, and a binary relation defining validity of Σ-formulas in Σ-models. Our framework is described in the same way, except that signatures are strongly inclusive (see definition 2). Moreover, we impose that the functor *Sen* maps all inclusion morphisms $\iota \in \mathcal{I}$ to the set-theoretic inclusion in *Set* between sets of formulas. More precisely, this condition means that if a signature is included in an other one, then the set of formulas constructed on the former is included in the set of formulas constructed on the latter. This constraint is natural. Besides, most of the formalisms that are widely used in computing science satisfy it. Thus, we have :

Definition 2. A *logic* is a quadruple $(Sig, Sen, Mod, \models)$ where:

- *Sig* is a strongly inclusive category whose associated inclusion system is $(\mathcal{I}, \mathcal{E})$. The objects of *Sig* are called *signatures*. Finally, *Sig* has an initial object, denoted $\Sigma_\emptyset$, which is also initial in $\mathcal{I}$.

- $Sen : Sig \rightarrow Set$ is a functor which for each $\Sigma \in |Sig|$ associates a set $Sen(\Sigma)$ of Σ-formulas. Moreover, Sen satisfies: $\forall(\Sigma \subseteq \Sigma')$, $Sen(\Sigma) \subseteq Sen(\Sigma')$;
- $Mod : Sig^{op} \rightarrow Cat$ is a contravariant functor which for each signature $\Sigma \in |Sig|$ associates a category $Mod(\Sigma)$ of Σ-models;
- $\models$ is a $|Sig|$-indexed family $\{\models_\Sigma\}_{\Sigma \in |Sig|}$ of relations $\models_\Sigma \subseteq Mod(\Sigma) \times Sen(\Sigma)$.

Let us now define specifications in the small as usual.

Definition 3. Given a logic $\mathcal{L} = (Sig, Sen, Mod, \models)$, a $\mathcal{L}$-specification SP is defined by a signature Σ and a set ax of formulas in $Sen(\Sigma)$. Given a $\mathcal{L}$-specification $SP = (\Sigma, ax)$, we note $Mod(SP)$ the full sub-category of $Mod(\Sigma)$ whose objects are all Σ-models $\mathcal{M}$ such that for any $\varphi \in ax$, $\mathcal{M} \models_\Sigma \varphi$. Finally, we note $SC(SP)$ the subset of $Sen(\Sigma)$, so-called *semantic consequences of SP*, defined as follows: $SC(SP) = \{\varphi \mid \forall \mathcal{M} \in |Mod(SP)|, \mathcal{M} \models_\Sigma \varphi\}$.

Within the framework of institutions, a supplementary condition, usually called the *satisfaction condition*, is added to formalisms. This condition expresses that properties are preserved through signature morphisms. That is to say : given a signature morphism $\sigma : \Sigma \rightarrow \Sigma'$, a Σ'-model $\mathcal{M}$ and a Σ-formula φ the following condition holds: $\mathcal{M} \models_{\Sigma'} Sen(\sigma)(\varphi) \Leftrightarrow Mod(\sigma)(\mathcal{M}) \models_\Sigma \varphi$. One of consequences of such a condition is that we can restrict the functor $Mod(\sigma)$ to the category of specifications. Indeed, given a specification morphism $\sigma : SP \rightarrow SP'$ (i.e. $\sigma : \Sigma \rightarrow \Sigma'$ and $Sen(\sigma)(ax) \subseteq ax'$), we have for any Σ'-models $\mathcal{M} \in Mod(SP')$ that $Mod(\sigma)(\mathcal{M}) \models_\Sigma ax$. Thus, we can restrict the functor $Mod(\sigma)$ from $Mod(SP')$ to $Mod(SP)$. This property is very useful for modularity purposes. Indeed, it ensures one of two ways of the modularity (so-called *modularity by restriction* in [4]) which expresses that a "complex" system can be used for simpler purposes [2]. However, one of the particularity of the feature paradigm is the ability to "break" the modularity. This ability is desired in order to keep a flexible conception. Indeed, a feature is built in isolation and not as an enrichment of the existing system on which it will be plugged on. A feature is only built as an enrichment of a subpart of this existing system (see Section 3). Thus, when a feature is plugged on the system, some unexpected behaviors (related to the existing system) can occur. This is one of the main preoccupation of feature-oriented system designers. This preoccupation is so-known as the *feature interaction problem*. Succinctly, when a new feature is plugged on a system, two kinds of perturbations can occur (as already mentioned in the introduction), which can respectively be represented by two types of logical properties :

---

[2] The other way is that a simple system can be extended to be reused for more "complex" purposes. This way is reached by imposing that there exists a synthesis functor (left adjoint to $Mod(\sigma)$) which is an inverse to the forgetful functor $Mod(\sigma)$ [4,9].

1. *Emerging properties:* the feature under analysis does more services than scheduled (by its specification) when it is plugged on the system.
2. *Lost properties:* the feature under analysis does less services than scheduled when it is plugged on the system.

The first kind of properties is due to the addition of new functionalities in the existing system. The second kind of properties might not be fully reached when the satisfaction condition holds (mainly for the last feature plugged on the system). Thus, imposing the satisfaction condition might prevent us to detect some interactions whose consequences would be harmful.

3   Feature over a logic

In this section, we formally define the notions of feature specification, feature integration, and feature interaction, given any logic.

3.1   Feature specifications

As we succinctly saw, a feature is a behavior defined as an extension of a required system. The required system is a subpart of the system on which we want to plug on the feature. Often, this subsystem is reduced to the original basic system. On this subsystem, we accept to modify any behavior except some invariants. These invariants are naturally semantic consequences of the original specification of the required system.

Definition 4. Let $\mathcal{L}$ be a logic. A $\mathcal{L}$-*feature* $\mathcal{F}$ is a triple (Req, Inv, SP) such that both $Req = (\Sigma, ax)$ and $SP = (\Sigma', ax')$ are $\mathcal{L}$-specifications, $\Sigma \subseteq \Sigma'$ and both $Inv \subseteq SC(Req)$ and $Inv \subseteq SC(SP)$. Req is called the *required specification* of the feature $\mathcal{F}$, Inv is called the set of *invariants* and SP is called its *feature specification*.

Then, the set Inv contains the invariants that we want to keep after the integration of the feature $\mathcal{F}$ as expressed by the condition $Inv \subseteq SC(SP)$. In this way, this last property can be seen as the feature correctness.

Thus, at the syntactical level, the building of a $\mathcal{L}$-feature is closely related to the notion of enrichment in the usual sense of algebraic specifications. Indeed, we can see a $\mathcal{L}$-feature as an "enrichment" of the specification (Σ, Inv) (i.e. defined in terms of a signature morphism together with a set of supplementary axioms [15]) except that we do not add supplementary axioms but only impose a conservation of the behavior (represented by Inv). On the contrary, at the model level, it will differ from the usual algebraic methods [4,9,15] in the sense that the feature is not required to be a conservative extension of the explicit behavior (represented by the required part) of the required system. Indeed, let us recall that in practice, integrating a feature often requires modifying the system where it will be plugged on. Thus, we cannot impose such modular constraints in order to ensure a separate design of the underlying existing system and the feature.

3.2   Feature systems and feature integration

Roughly speaking, you can see a feature system Sys as a set of features provided with a structuration denoting that each feature is described on a subset of Sys. The key question is how to define feature-oriented systems provided with an adequate semantic counterpart. A first answer might be to consider the union of all axioms issued from the different features of the system under analysis. However, this is not the good solution. Indeed, let us recall that each feature is defined as possibly modifying the behavior of the existing system (more precisely, a subpart of it) where it will be plugged on. Thus, any system obtained would be lucky enough to be inconsistent. The good answer is to provide the system with an order of integration compatible with requirements of each feature. Thus, a feature $\mathcal{F}_2$ has to be plugged after a feature $\mathcal{F}_1$ if its required system depend on $\mathcal{F}_1$. So, for two independent features this order is arbitrary. Syntactically, a feature system is then defined as follows:

Definition 5. Let $\mathcal{L}$ be a logic. A $\mathcal{L}$-feature system Sys is a finite sequence of different features $(\mathcal{F}_i)_{i \leq n}$ such that $\mathcal{F}_1 = ((\Sigma_\emptyset, \emptyset), \emptyset, SP_1)$ and is called the basic system.

Remark 2. Definition 5 does not suggest that features are known in advance. Such a system only describes the already designed features.

Models do not take into account structurations. Thus, the semantical counterpart of a $\mathcal{L}$-feature system Sys (i.e. the category of models validating it) has to be defined from a flat specification. This flat specification has to be the result of successive integrations of the features of Sys, respecting the order of integration. However, this is not sufficient. Indeed, let us recall that a feature is defined in isolation. Thus in practice, between the description of any feature and its integration, other features will probably be defined and plugged on the existing system. Moreover, among those features, some of them may be defined on the same required specification. Thus, when we integrate any feature, its required specification may no longer exist. Nevertheless, this required specification has to occur at least in the whole integration history of the system existing before the integration of the feature of interest. When this constraint is satisfied by all the features under consideration, we will say that the system is *well-designed*. Let us formally define this notion :

Definition 6. Let $\mathcal{L}$ be a logic. Let $Sys = (\mathcal{F}_i)_{i \leq n}$ be a $\mathcal{L}$-feature system. Sys is said to be *well-designed* if and only if we can inductively define a finite sequence of specifications $(Sys_i)_{i \leq n}$, called *feature-integration-sequence of Sys*, as follows :

- $Sys_1 = SP_1$;

- for $\mathcal{F}_{k+1} = (Req_{k+1}, Inv_{k+1}, SP_{k+1})$ with $Req_{k+1} = (\Sigma_{k+1}, ax_{k+1})$ and $SP_{k+1} = (\Sigma'_{k+1}, ax'_{k+1})$, there exists $j \leq k$ such that $Req_{k+1} \subseteq Sys_j$ (i.e. $\Sigma_{k+1} \subseteq \Sigma_j^{Sys}$ and $ax_{k+1} \subseteq ax_j^{Sys}$), and

$$Sys_{k+1} = (\Sigma_k^{Sys} + \Sigma'_{k+1}, (ax_k^{Sys} \setminus ax_{k+1}) \cup ax'_{k+1})$$

For the sake of the paper, all $\mathcal{L}$-*feature system* will be considered as *well-designed*.

Definition 6 calls some comments :

- It may seem strange to integrate a feature relying on some properties that have been removed. However it is what happens in a real feature-oriented system design. Indeed, when a feature is intended to modify some behaviors of some already existing modules, then nothing ensures that some others modifications have been or will be done on those modules when integrating some others feature. Even-though our modelisation decisions about the designing and the integration of a feature seem to fit with reality, several real-size case studies will be useful to validate them and see how far they are practical.
- Intuitively, Sys_{k+1} denotes the integration of the feature $\mathcal{F}_{k+1}$ on $Sys_k = (\Sigma_k^{Sys}, ax_k^{Sys})$. Roughly speaking, the behavior of the system that the feature $\mathcal{F}_{k+1}$ aims to modify (i.e. ax_{k+1}) is replaced by the new behavior specified in $\mathcal{F}_{k+1}$ (i.e. ax'_{k+1}), in the global behavior of the system (i.e. ax_k^{Sys}). In some way, the proposal simply boils down to replace a subset of axioms of the basic system by an other one. In fact, our abstract vision of the integration problem is relatively close to the one described in [16], that is to say, plugging a module on the system (modify Sys_k to Sys_{k+1}), which declares what module(s) of the system, the feature needs in order to make sense (ax_{k+1}) and expresses the modification to be done on this module (ax'_{k+1}).

Thus, Sys_n is the final flat specification of the $\mathcal{L}$-feature system and has to be understood as the "final system" obtained after integration.

Consequently, at the semantic level, the mathematical meaning of Sys is simply defined as :

Definition 7. With the previous notations, given a *well-designed* $\mathcal{L}$-feature system $Sys = (\mathcal{F}_i)_{i \leq n}$, then :

$$Mod(Sys) = \{\mathcal{M} \in |Mod(Sys_n)| \backslash \forall i \in \{1 \cdots, n\}, \mathcal{M} \models Inv_i\}$$

Then, a feature system is consistent if there is at least one (untrivial[3]) model in $Mod(Sys)$ which satisfies it. Thus, feature integration problems are reduced to verify that feature systems obtained after integration are consistent.

---

[3] i.e. a model which satisfies all formulas, as the terminal model does for the usual conditional equational logic.

3.3   Feature interactions

As we glimpsed in Section 2, the feature interaction problem is strictly speaking a "break" of the modularity. Thus, the feature interaction problem characterises unexpected behaviors of a feature when it is plugged on a system. In this way, the feature interaction problem consists in comparing a feature specification with respect to the system specification where it has been plugged on.

To simplify, those perturbations can be classified in two large classes when comparing a system S with any feature already integrated on it [4] :

1. those denoted by properties dealing with the feature of interest, which are true for the system (due to the integration of some other features) but not for the feature description. We will call such a property : "emerging property".
2. those denoted by properties dealing with the feature of interest, which are true for the feature description but which do not hold anymore for the system (due to the integration of some other features). We will call such a property : "lost property".

We will say that a "feature interaction" occurs between a feature and the system on which it has been integrated, when there exists some perturbations as described above. More formally, feature interactions can be expressed as follows :

Definition 8. Let $Sys = (\mathcal{F}_i)_{i \leq n}$ be a feature system. For each $i \in \{1, \ldots, n\}$, let us call:

1. *Set of emerging properties* for the feature $\mathcal{F}_i$ w.r.t. the system Sys the set EP_i defined by: $EP_i = SC(Sys) \setminus SC(SP_i)$;
2. *Set of lost properties* for the feature $\mathcal{F}_i$ w.r.t. the system Sys the set LP_i defined by: $LP_i = SC(SP_i) \setminus SC(Sys)_{\restriction \Sigma_i}$.

   where SP_i denotes the specification of $\mathcal{F}_i$ and
   $$SC(Sys)_{\restriction \Sigma_i} = \{\varphi \in Sen(\Sigma_i) \mid \varphi \in SC(Sys)\}$$
   (by Definition 2, $\Sigma_i \subseteq \sum_{j=1}^{n} \Sigma_j$ so $Sen(\Sigma_i) \subseteq Sen(\sum_{j=1}^{n} \Sigma_j)$).

   Then we will say that a feature interaction occurs between $\mathcal{F}_i$ and Sys if and only if either EP_i or LP_i is non-empty.

Remark 3. Of course, because feature conception does not follow a modular approach, any addition of a new feature $\mathcal{F}$ in a system Sys requires to check

---

[4] In the literature, there exist several taxonomies, according to the different points of view followed by the authors (see [14] for a survey). However, all the interactions they classify can be defined in our framework.

again the set described above for all features in $Sys + \mathcal{F}$ (i.e. both for $\mathcal{F}$ and for all features $\mathcal{F}'$ in Sys). This problem has already been raised with the object-oriented approach when dealing with concurrent (active) object systems as this has been shown in [1].

4   An example of a logic

We now introduce a particular instance of logics. The logical framework presented in this section is an extension of the usual many-sorted equational logic to dynamic aspects. It is a restriction of more complete formalisms initially developed to deal with concurrent (active) object systems [1] and afterwards to deal with real-time systems [3]. The main idea of this formalism is to extend algebraic approaches where the dynamic behavior of object systems relies on implicit states used as modifiers of semantics of functionalities. This formalism can be related to works mainly influenced by the Gurevich's evolving algebra approach [2,13] except states are simple values but not anymore algebras. At least, it can also be related to the Goguen's hidden algebras [11]. Below, we define its four components $(Sig, Sen, Mod, \models)$.

4.1   The category Sig of signatures

Objects and morphisms of Sig are respectively defined by:

Definition 9. A *signature* Σ is a quadruple (S, F, F_{obs}, V) where S is a set of sort names, F is a set of function names each one provided with an arity of the form $(s_1 \times \ldots \times s_n \to s)$ such that each $s_i \in S$ with $i \in \{1, \ldots, n\}$ and $s \in S \cup \{\varepsilon\}$ (ε denotes the empty sort and will be used to characterise functions whose the behavior is only to modify states), F_{obs} is a subset of F such that for all $(f : s_1 \times \ldots \times s_n \to s) \in F_{obs}$, $s \in S$ (F_{obs} denotes the subset of functions names that will be used to observe states), and V is a S-indexed family of sets of variables $\{V_s\}_{s \in S}$.

Definition 10. With the previous notations, morphisms are all arrows $\sigma :$ $(S_1, F_1, F_{1\,obs}, V_1) \to (S_2, F_2, F_{2\,obs}, V_2)$ defined by three applications $\sigma_1 :$ $S_1 \to S_2$, $\sigma_2 : F_1 \to F_2$, and $\sigma_2 : V_1 \to V_2$ such that σ_2 respects function arity, $\sigma_2(F_{1\,obs}) \subseteq F_{2\,obs}$, and σ_3 is injective.

Thus, a morphism $\sigma = (\sigma_1, \sigma_2, \sigma_3)$ is said to be an *inclusion* if each one of these three applications denote set inclusions, to be *injective* if each one of these three applications are injective, and to be *surjective* if σ_1, σ_2 and σ_3 are surjective (and thus σ_3 is bijective) and $\sigma_2(F_{1\,obs}) = F_{2\,obs}$. The inclusion system for Sig is then defined by the pair $(\mathcal{I}, \mathcal{E})$ where $\mathcal{I}$ contains all inclusions between signatures and $\mathcal{E}$ contains all surjections. Trivially, Sig is strongly inclusive and has as initial object the signature $\Sigma_\emptyset = (\emptyset, \emptyset, \emptyset, \emptyset)$.

4.2   The functor *Sen*

Sen is the functor which from a signature yields the set of the well-formed formulas. Thus, given a signature Σ we obtain the following definition for $Sen(\Sigma)$:

Definition 11. Given a signature $\Sigma = (S, F, F_{obs}, V)$, $Sen(\Sigma)$ denotes the least set of strings of symbols so that:

- for every pair of terms with variables [5] of the same sort in S, (t, t'), the equation $t = t'$ is in $Sen(\Sigma)$ (such a formula denotes an equation between values);
- for every pair of terms with variables of the same sort in $S \cup \{\varepsilon\}$ (t, t'), the equation $t \equiv t'$ is in $Sen(\Sigma)$ (such a formula denotes an equation between states);
- for every formula $\varphi \in Sen(\Sigma)$ and every term with variables t, the formula **after** $[t]$ (φ) is in $Sen(\Sigma)$ (succinctly, such a formula means that φ has to be true just after that t has been performed);
- is closed under usual first-order logic connectives and quantifiers.

Finally, from a morphism signature $\sigma : \Sigma_1 \to \Sigma_2$, $Sen(\sigma)$ is its canonical extension $\sigma^{\#}$ to formulas. Trivially, the functor *Sen* respects inclusions in *Sig* (i.e. if $\Sigma_1 \subseteq \Sigma_2$ then $Sen(\Sigma_1) \subseteq Sen(\Sigma_2)$).

4.3   The functor *Mod*

As usual, given a signature Σ we associate the category of Σ-models, so-noted $Mod(\Sigma)$. Σ-models and Σ-morphisms are defined as follows:

Definition 12. Given a signature Σ, a Σ-*model* $\mathcal{A}$ is a pair $(A, \underline{A})$ where A, called the *set of values*, is a S-indexed family of sets $\{A_s\}_{s \in S}$ and $\underline{A}$, called the *set of states*, is a set, together for each function $f : s_1 \times \ldots \times s_n \to s$ in F with two $\underline{A}$-indexed family of applications :

1. $\{f_\eta^{\mathcal{A}} : A_{s_1} \times \ldots \times A_{s_n} \to A_s\}_{\eta \in \underline{A}}$ with as convention that $A_\varepsilon = \mathbf{1}$ ($\mathbf{1}$ is the neutral element for the Cartesian product of sets);
2. $\{\underline{f}_\eta^{\mathcal{A}} : A_{s_1} \times \ldots \times A_{s_n} \to \underline{A}\}_{\eta \in \underline{A}}$.

Definition 13. Given two Σ-models $\mathcal{A}$ and $\mathcal{B}$, a morphism $\mu : \mathcal{A} \to \mathcal{B}$ is a pair of applications $(\mu, \underline{\mu})$ such that $\mu : A \to B$ respects sorts (i.e. $\mu(A_s) \subseteq B_s$) and $\underline{\mu} : \underline{A} \to \underline{B}$ and such both μ and $\underline{\mu}$ are compatible with functions (i.e. both $\mu(f_\eta^{\mathcal{A}}(a_1, \ldots, a_n)) = f_{\underline{\mu}(\eta)}^{\mathcal{B}}(\mu(a_1), \ldots, \mu(a_n))$ and $\underline{\mu}(\underline{f}_\eta^{\mathcal{A}}(a_1, \ldots, a_n)) = \underline{f}_{\underline{\mu}(\eta)}^{\mathcal{B}}(\mu(a_1), \ldots, \mu(a_n))$).

---

[5] The set of terms with variables is the $S \cup \{\varepsilon\}$-family of set $\{T_\Sigma(V)_s\}_{s \in S \cup \{\varepsilon\}}$ where for each $s \in S \cup \{\varepsilon\}$, $T_\Sigma(V)_s$ contains all the s-sorted terms with variables that are inductively built from the set of variables V_s and using function symbols of F.

We now define the *forgetful functor* $Mod(\sigma)$ as usual:

Definition 14. Given a signature morphism $\sigma : \Sigma_1 \rightarrow \Sigma_2$, $Mod(\sigma) : Mod(\Sigma_2) \rightarrow Mod(\Sigma_1)$ is defined as follows:

- given a Σ_2-model $\mathcal{A}$, $Mod(\sigma)(\mathcal{A})$ is the Σ_1-model $\mathcal{B}$ defined by the pair $(B, \underline{B})$ where $B = \{A_{\sigma(s)}\}_{s \in S_1}$ and $\underline{B} = \underline{A}$ and for each η in B and for each f in F_1, $f_\eta^\mathcal{B}$ (resp. $\underline{f}_\eta^\mathcal{B}$) is $\sigma(f)_\eta^\mathcal{A}$ (resp. $\underline{\sigma(f)}_\eta^\mathcal{A}$).
- given a Σ_2-morphism $\mu : \mathcal{A} \rightarrow \mathcal{B}$, $Mod(\sigma)(\mu)$ is the Σ_1-morphism μ' defined by all the restrictions of the form: $\mu_s' = \mu_{\sigma(s)}$.

4.4   The satisfaction relation

With any extension of first order logic, defining the satisfaction relation requires at first to evaluate terms. Unlike usual algebraic approaches we cannot canonically extend any assignment ν in an application $\nu^{\#} : T_\Sigma(V) \rightarrow A$. This is due to side-effects of any sub-terms of a term t. Consequently, the order where sub-terms are evaluated is important. As we do not want to impose any evaluation order, it results in a set of possible values and states.

Definition 15. The *evaluation* of a term $t = f(t_1, \ldots, t_n)$ of $T_\Sigma(A)$ from $\eta \in \underline{A}$ is defined as follows:

- if for all $i \in \{1, \ldots, n\}$ $t_i \in A$ (in this case t is called a *flat term*) then the evaluation of t is the pair $(f_\eta^\mathcal{A}(t_1, \ldots, t_n), \underline{f}_\eta^\mathcal{A}(t_1, \ldots, t_n))$
- otherwise, let t' be any flat sub-term of t and (a', η') be the result of its evaluation. The pair (t'', η') where t'' is the term obtained form t by replacing the occurrence of the flat sub-term t' by a' is called a *partial evaluation* of t. Let us note $\mathcal{P}$ the binary relation on $T_\Sigma(A) \times \underline{A}$ to denote this partial evaluation. Consequently, the evaluation of t from η is the relation $\mathcal{K} \subseteq (T_\Sigma(A) \times \underline{A}) \times (A \times \underline{A})$ defined as the reflexive and transitive closure of $\mathcal{P}$ restricted to values in A.

It is well-known that in practice it is often impossible to completely observe the states of a software due to implementation details. To give an algebraic treatment of this problem, we use behavioral approach to satisfy equations between states [6]. As usual, the behavioral satisfaction is defined from the notion of observable contexts.

Definition 16. Given a signature $\Sigma = (S, F, F_{obs}, V)$, a Σ-*context* is a term c which is of the form $f(t_1, \ldots, t_n)$ with $(f : s_1 \times \ldots \times s_n \rightarrow s) \in F_{obs}$, having a *single* variable x and a *single* occurrence of this variable. Usually, it is noted $c[x]$. Given a Σ-context $c[x]$, let us note $c[x] : s$ to mean that the variable x is of sort s. For any term t of sort s, $c[t]$ is the term obtained from $c[x]$ by replacing the variable x by t.

---

[6] This idea is not new and has already been used in [12].

We are in position to define our satisfaction relation.

Definition 17. Given $\mathcal{A} \in Mod(\Sigma)$ and $\varphi \in Sen(\Sigma)$, $\mathcal{A}$ *satisfies* φ for an assignment $\nu : V \to A$ [7] and a state $\eta \in \underline{A}$, denoted $\mathcal{A} \models_{\nu,\eta} \varphi$, if and only if:

- if φ is of the form $t = t'$ then $\mathcal{A} \models_{\nu,\eta} \varphi$ if and only if for every $(t, \eta)\mathcal{K}(a, \eta_1)$ and every $(t', \eta)\mathcal{K}(a', \eta_2)$ we have: $a = a'$;
- if φ is of the form $t \equiv t'$ then $\mathcal{A} \models_{\nu,\eta} \varphi$ if and only if for every Σ-context $c[x]$, $\mathcal{A} \models_{\nu,\eta} c[t] = c[t']$;
- if φ is of the form **after** $[t]$ (ψ) then $\mathcal{A} \models_{\nu,\eta} \varphi$ if and only if for every $(t, \eta)\mathcal{K}(a, \eta')$, $\mathcal{A} \models_{\nu,\eta'} \psi$;
- the others connectives and quantifiers are handled as usual.

$\mathcal{A} \models_{\Sigma} \varphi$ if and only if for every $\nu : V \to A$ and every $\eta \in \underline{A}$ $\mathcal{A} \models_{\nu,\eta} \varphi$

Thus, $\models$ is the $|Sig|$-family $\{\models_{\Sigma}\}_{\Sigma \in |Sig|}$. From such a definition of satisfaction, it is easy to see that the satisfaction condition does not hold in general. Indeed, given a signature morphism $\sigma : \Sigma_1 \to \Sigma_2$, to behaviorally satisfy Σ_1-equations between states we have to prune all contexts having a term $f(t_1, \ldots, t_n)$ with $f \in F_{2_{obs}} \setminus \sigma(F_{1_{obs}})$ as sub-term. Of course, imposing the restriction on signature morphisms to forbid to add new functionalities provided with a sort s of the old signature as co-domain, allows to retrieve this property. However, one typical feature of the feature paradigm is that a feature adds new services (i.e. new functionalities) by modifying existing ones.

5 Examples of feature interactions

We now aim to illustrate both concepts of *emerging properties* and *lost properties* in presence of two features. Although we said that features-oriented systems are a new paradigm which goes further than software telecommunication systems field, our example also aims to show that our characterisation of the notion of interaction is relevant regarding the existing type of interactions in a software telecommunication systems (where those concepts have emerged). Hence, we give a simple specification of a feature-oriented telecommunication system. In order to facilitate the perception of these concepts, our specification is written at a high-level of description. All specifications will be described both using the general presentation given in Section 3 and using the formalism described in Section 4 as particular instantiation. Those two features will be described on the same required specification issued from $\mathcal{F}_{POTS}$ which represent intuitively the basic system. Intuitively, it describes the ability for two given phones to be engaged together. The informal specifications of both features $\mathcal{F}_{CFU}$ and $\mathcal{F}_{RC}$ are respectively :

---

[7] An assignment $\nu : V \to A$ is a S-indexed family of applications $\{\nu_s : V_s \to A_s\}_{s \in S}$.

- *CFU* (call forward unconditionnaly) allows a subscriber to re-send unconditionnaly all the calls that his phone receives to another phone.
- *RC* (restrained-call) allows to forbid all communications involving two phones which are not in the same geographical area.

$\mathcal{F}_{POTS}$: Since $\mathcal{F}_{POTS}$ aims at describing the ability for two phones to be engaged together, thus $\mathcal{F}_{POTS}$ is defined with an empty required specification $(\Sigma_\emptyset, \emptyset)$, an empty set of invariants Inv. Finally, the feature specification $SP_r = (\Sigma_r, ax_r)$ is given by:

- $\Sigma_r = (S_r, F_r, F_{r_{obs}}, V_r)$ is the signature where :
  $S_r = \{bool, phone\}$.
  $$F_r = \left\{ \begin{array}{l} dial : phone \times phone \to \varepsilon \\ on\_line? : phone \times phone \to bool \\ create\_line? : phone \to bool \\ free : phone \to bool \end{array} \right\}$$
  $F_{r_{obs}} = \{create\_line? : phone \to bool\}$
  V_r is the set of the Latin small letters
- ax_r consists of the following axioms :

  ax_1 : $\neg(x = y) \Rightarrow$
  $\qquad$ **after** $[dial(x, y)]$ $(create\_line?(x) = true \land free(x) = false)$

  Whatever the state of the system is, after x has dialled y's number, x is in a state in which a he is asking for being on line with y. At our level of description, we do not precise the target and just express that x can create a line, which is meant by : $create\_line?(x) = true$. Of course, after x has dialled y's number, x is busy, which is meant by : $free(x) = false$.

  ax_2 : $\neg(x = y) \Rightarrow$
  $\qquad$ **after** $[dial(x, y)]$ $((x \equiv y \land free(y) = true) \Leftrightarrow on\_line?(x, y) = true)$

  By following our behavioral satisfaction, after x has dialled y, if we observe that y is in the same state than x (i.e. $create\_line?(y) = true$, cf : ax_1 and $F_{r_{obs}}$), then if y is not busy (i.e. free(y)=true), a line between x and y can be created. Of course, the effective creation of this line depends on the fact that y takes the hook off, but this is not expressed in our specification.

Let us note that we do not impose $free(x) = true \Rightarrow create\_line?(x) = true$. Indeed, $free(x) = true$ only means that the phone is on the hook. However, for many reasons (as for example the bill is not paid), it is possible $create\_line?(x) = false$.

The C.F.U. feature : The C.F.U. feature $\mathcal{F}_{CFU}$ is defined upon the required specification SP_r. It is defined as a restriction of the axiom 2 above. Indeed, if we observe that the phone y has performed a C.F.U. on a phone z, then, after having dialled the phone-number of y, x will be able to be engaged with z, but not with y anymore. Thus, we describe the C.F.U. feature with SP_r as required specification and with $\{ax_1\}$ as the set of invariants. The feature specification SP_{CFU} of C.F.U. is composed of :

- $\Sigma_{CFU} = (S_{CFU}, F_{CFU}, V_{CFU})$ is the signature where :
  $S_{CFU} = S_r$
  $F_{CFU} = F_r \cup \{cfu : phone \times phone \rightarrow bool\}$
  $F_{CFU\,obs} = \{create\_line? : phone \rightarrow bool\}$
  V_{CFU} is the set of the Latin small letters
- The set of axioms ax_{CFU} consists of the first axiom of ax_r together with the two following axioms :

ax_3 : $\neg(x = y) \Rightarrow$
$\quad\quad\quad$ **after** $[dial(x,y)]$ $(\exists z(cfu(y,z) = true) \Rightarrow$
$\quad\quad\quad\quad\quad\quad\quad\quad ((x \equiv z \wedge free(z) = true) \Leftrightarrow on\_line?(x,z) = true))$

The axiom ax_3 expresses that if y has forwarded its calls on the phone z then, when x dials the phone-number of y, all seems as if x has dialled the phone-number of z. Note that, in our description, nothing avoids a phone to have a cfu on itself. In this case, all happens as if there was no forward. Moreover, we do not express the fact that a cfu is unique. Note that if $z = x$, then $on\_line?(x,z) = false$ because $free(x) = false$ (see ax_1).

ax_4 : $\neg(x = y) \Rightarrow$
$\quad\quad\quad$ **after** $[dial(x,y)]$ $(\forall z(cfu(y,z) = false) \Rightarrow$
$\quad\quad\quad\quad\quad\quad\quad\quad ((x \equiv y \wedge free(y) = true) \Leftrightarrow on\_line?(x,y) = true))$

The axiom ax_4 expresses that when no cfu has been performed by y, the side-effect of $dial(x,y)$ is the same than in SP_r.

The R.C. feature : The R.C. feature $\mathcal{F}_{RC}$ is also defined upon SP_r. As the feature $\mathcal{F}_{CFU}$, this feature is defined as a restriction of ax_2. Thus $\mathcal{F}_{RC}$ is given by SP_r as required specification, $\{ax_1\}$ as set of invariants, and the feature specification SP_{RC} composed of :

- $\Sigma_{RC} = (S_{RC}, F_{RC}, V_{RC})$ is the signature where :
  $S_{RC} = S_r \cup \{crit\}$.
  $F_{RC} = F_r \cup \{area : phone \rightarrow crit, rc : phone \rightarrow bool\}$.
  $F_{RC\,obs} = \{create\_line? : phone \rightarrow bool, area : phone \rightarrow crit\}$
  V_{RC} is the set of the Latin small letters
  Intuitively, $area : phone \rightarrow crit$ denotes a functionality yielding some

informations about the area the phone is situated in. Thus, $area(x) = area(y)$ means that both phones x and y are in the same area.

- The set of axioms ax_{RC} consists of the first axiom of ax_r together with the two following axioms :

ax_5 : $\neg(x = y) \Rightarrow$
    after $[dial(x,y)]$ $(rc(x) = true \Rightarrow$
        $((x \equiv y \wedge free(y) = true) \Leftrightarrow on\_line?(x,y) = true))$

ax_6 : $\neg(x = y) \Rightarrow$
    after $[dial(x,y)]$ $(rc(x) = false \Rightarrow$
        $((create\_line?(y) = true \wedge free(y) = true) \Leftrightarrow on\_line?(x,y) = true))$

Let us consider the feature system $Sys = \{\mathcal{F}_{POTS}, \mathcal{F}_{CFU}, \mathcal{F}_{RC}\}$. According to the Definition 8, its corresponding flat specification is given by $SP_{Sys} = (\Sigma_{POTS} + \Sigma_{CFU} + \Sigma_{RC}, \{ax_1, ax_3, ax_4, ax_5, ax_6\})$.

Lost property. Let φ be the following formula :
$\neg(x = y) \Rightarrow$
    after $[dial(x,y)]$ $(\exists z(cfu(y,z) = true) \Rightarrow$
        $((create\_line?(z) = true \wedge free(z) = true) \Leftrightarrow on\_line?(x,z) = true))$

From the two additional axiom ax_3 described in SP_{CFU}, it is easy to see that φ belongs to $SC(SP_{CFU})$. Now, this property is not kept by the system Sys. Indeed, $x \equiv z$ now also requires that $area(x) = area(z)$. Thus φ is a lost property of the feature $\mathcal{F}_{CFU}$ with respect to Sys. This property exhibits an interaction between $\mathcal{F}_{CFU}$ and Sys, due to the integration of $\mathcal{F}_{RC}$, which increases observations.

Emerging property. In the system Sys, we have the following formula as consequence:

$\neg(x = y) \Rightarrow$
after$[dial(x,y)]$
$((\exists z(cfu(y,z) = true \wedge rc(x) = true \wedge \neg(area(x) = area(z))))) \Rightarrow$
$on\_line?(x,z) = false)$

This emerging property raises an interaction between both $\mathcal{F}_{RC}$ and $\mathcal{F}_{CFU}$, and Sys. This come from the fact that we do not treat any precedence rules for the appliance of the two features.

6   Conclusion

In this paper, we have introduced a general framework to specify any system in a feature-oriented style. In this way, features have been defined by a set of invariants of the existing system that the feature ensures to preserve, and a set of properties specifying the feature. Feature-oriented systems are defined by an incremental process of feature integration. An interactions between a feature and a system on which the feature has been integrated occurs when one of the sets of *emerging properties* or *lost properties* for the feature with respect to the system is non-empty.

Emerging properties denote behaviors in addition than scheduled by the specification of the feature when lost properties denote behaviors in less.

We intend to pursue this work by introducing the notion of scenario within our general framework. We think that, with such a notion, we could give feature interactions under a canonical form based on a scenario leading to an observable result. This would be useful for telecommunication experts to analyse the qualitative nature of the detected interactions. Above all, we want to take benefits of our algebraic setting. At present, we are investigating towards refinement or testing algebraic techniques to be applied to our logics.

7   Acknowledgements

We especially thank Gilles Bernot for constructive comments on the ideas developed in this paper. we also thank Mark Ryan, Pierre-yves Schobbens and Franck Cassez for all fruitful discussions.

Finally we also would like to thank the anonymous referees of this paper, who made plenty of useful comments. Those comments really helped us to increase the quality of the paper.

References

1. M. Aiguier and G. Bernot. *Information Systems - Correctness and Reusability*, chapter Algebraic semantics of object type specifications, pages 16–32. World Scientific, 1995. Selected Papers from the IS-CORE Workshop.
2. E. Astesiano and E. Zucca. D-oids: A model for dynamic data types. *Mathematical Structure in Computer Science*, 1994. Special Issue.
3. S. Béroff. *Présentation d'un formalisme de spécification dynamique et temps réel ; application à la sémantique du langage VHDL*. PhD thesis, University of Evry, january 1999.
4. M. Bidoit. The stratified loose approach. a generalization of initial and loose semantics. In *Recent Trends in Data Type Specification*, volume 332 of *LNCS*, pages 1–22. Springer-Verlag, 1988.
5. M. Bidoit, D. Sannella, and A. Tarlecki. Architectural specification in CASL. In *AMAST'98, Amazonia-Manaus*, volume to appear. Springer, LNCS, 1999.
6. L.G. Bouma and H. Velthuijsen, editors. *Feature Interactions in Telecommunications and Software Systems (FIW'95)*. IOS Press, 1995.
7. L.G. Bouma and H. Velthuijsen, editors. *Feature Interactions in Telecommunications and Software Systems (FIW'98)*. IOS Press, 1998.
8. M. Calder. What use are formal design and analysis methods to telecommunications services. In *[7]*, pages 23–31, 1998.
9. R. Diaconescu, J. Goguen, and P. Stefaneas. Logical support for modularization. In G. Huet and G. Plotkin, editors, *Logical Environments*, Proceedings of a Workshop on Logical Frameworks, pages 83–130, may 1991.

10. J. A. Goguen and R. M. Burstall. Institutions: abstract model theory for specifications and programming. *association for Computing Machinery*, 1992.
11. J.A. Goguen and R. Diaconescu. Towards an algebraic semantics for the object paradigm. In *LNCS 785*, volume 785, pages 1–29, 1994.
12. J.A. Goguen and G. Malcom. Hidden agenda. *Theoritical Computer Science*, 1999.
13. Y. Gurevich. Evolving algebras, an attempt to discovery semantics. In G. Rozenberg and A. Salomaa, editors, *Recent Trends in Technical Computer Science*. World Scientific, 1993.
14. D. O. Keck and J. Kuehn. The feature and service interaction problem in telecommunication systems : A survey. *IEEE Transactions on Software Engineering*, 24(10):779–796, october 1998.
15. F. Orejas. *Algebraic Foundations of Systems Specifications*, chapter Structuring and Modularity, pages 159–200. IFIP State-of-the-Art Reports. Springer, 1999.
16. M. Plath and M. Ryan. Plug-and-play features. In *[7]*, pages 150–164, 1998.
17. A. Sernadas, C. Sernadas, and C. Caleiro. Denotational semantics of object specification. *Acta Informatica*, 35:729–773, 1998.

The PEPA Feature Construct

Stephen Gilmore and Jane Hillston

LFCS, The University of Edinburgh, Edinburgh EH9 3JZ, Scotland.
Email: {stg, jeh}@dcs.ed.ac.uk

Abstract. We show how the PEPA performance modelling language could be extended with a *feature construct* which can be used to describe modifications to PEPA models. We provide this construct with an operational description which conservatively extends the operational semantics of the PEPA language. We then show how the feature construct can be applied in a small case study.

1 Introduction

PEPA (Performance Evaluation Process Algebra) [1] is a performance modelling notation. It is also a process algebra, a concise mathematical language which is amenable to formal reasoning. PEPA is defined by an unambiguous semantics which makes clear the meaning of all models which are expressed in the language. It has been used to investigate the behaviour and performance of a diversity of distributed and concurrent systems [2–6].

Constructing performance models of distributed systems is a worthwhile activity. Distributed systems often have designs which are both complex and novel. An ill-considered design decision can lead to an implementation which fails to achieve planned levels of service or has unnecessarily high running costs.

As is the case for other performance modelling notations, PEPA can be applied either prospectively to assess the viability of a candidate design for a yet-to-be-constructed system or retrospectively to provide insight into a fully functional operation. Our novel contribution in this paper is to show how the PEPA language could be extended with syntactic support for the description of enhancements to models which reflect enhancements to the system under study. We use the term *features* to describe these enhancements to both systems and models, in keeping with the use of this term in the telecommunications industry and in software development. With this extension, the PEPA language can more easily be used throughout the entire lifetime of a complex distributed system, tracking adaptive and corrective maintenance in a formal setting.

Improving the suitability of the PEPA language for modelling complex systems is a useful extension. The difficulty in computer system development stems from the desire to create sophisticated and comprehensive products. These are constructed piece by piece and are subject to many unpredictable revisions over time. This is our motivation for considering a formal means of expressing the addition of new features to an exisiting system. The fact that a system has interesting performance qualities which are worth investigating via the construction and analysis of system models does not make it impervious to modifications and the addition of new features.

One way to address the problem of modelling the addition of features is to modify the extant system description in such a way that the new feature is incorporated or interwoven into the model description. This might at first seem to be an attractive option since it does not require any modification to the existing modelling language which was used to describe the original system. The modeller can simply think that they are only constructing a model of a more complex system, namely the one which includes the added feature. However, this approach has the disadvantage that it reduces the intellectual leverage which the identification of features gives both to system designers and system builders. Firstly, the valuable documentation function which features provide has not been exploited with the consequence that no formal record of the change history of the system is being created. Secondly, the dependency of the added feature upon components of the existing system will be unclear. Effectively, *all* of the existing system description has been considered to be essential to the description of the new feature. It is not usual that this is the case in practice.

We adopt the principle that feature descriptions require a different form of expression from system descriptions. Particularly, their descriptions should make clear the dependencies of components of the existing system. This places a demand upon our modelling language to provide some distinctive syntactic support for the formal expression of features. Other authors have also argued that the addition of a feature construct to an existing modelling language is the right method by which to make progress in this problem [7,8].

2   Design of the feature construct

PEPA is a small language with essential, simple combinators. A description appears in Appendix A. Readers who are already familiar with the PEPA language can omit this summary, which is standard. Readers unfamiliar with PEPA who are keen to understand the technical details of this paper should study the language summary in Appendix A before proceeding. Briefly, PEPA *components* perform *timed activities*. Each PEPA model defines a labelled multi-transition system which can be read as a Continuous Time Markov Chain (CTMC) by ignoring the activity names which label the arcs from one state to another.

A feature construct for PEPA must add value without incurring unnecessary loss of simplicity. (The feature construct can itself be seen as a *feature* which is being added to the existing PEPA language. Many of the good practices which are applied when adding functional features to software systems have their analogues here where we are adding a model structuring feature to an existing modelling language.)

Following [7], we consider that a feature construct should describe features formally as self-contained units of functionality. It should be possible to consider features in isolation, without complete knowledge of the system to which they are being added. A feature construct should be general-purpose, allowing a number of different types of features to be added. However, it should not allow undisciplined modifications which would be hard to reason about or understand. This last requirement would rule out of consideration as candidates a number of powerful, but low-level, macro-like operators.

In the particular setting of the PEPA modelling language we want a feature construct which is applicable, general, clear and easy to explain. It must have a formal definition. Two candidates present themselves as being possibly suitable; *re-binding* and *parametric definition*.

2.1   Re-binding component definitions

PEPA is a *compositional* description language so in principle new features could be installed by re-binding the definitions of key components. Such an extension to the PEPA language would meet most of our criteria for a feature construct. Re-binding definitions is a general-purpose concept and it can certainly be described both formally and with clarity. Unfortunately, it fails to meet our key criteria of applicability because it cannot describe the most general case of making unforeseen extensions to an existing system. The use of re-binding as a feature construct is only applicable in the cases where the designers of the system have previously loaded the system with re-programmable hooks. It is usual to describe such systems as *feature-ready* because they have been designed in the anticipation of the addition of new features in particular ways. Extensible software systems such as Web browsers with a "plug-in" capability are an example of feature-ready systems.

2.2   Parametric components

Our winning candidate for a feature construct for PEPA is the use of parameterised components. The parameters capture the dependency of the feature on the existing system. By defining our new feature in terms of the existing system, monitored for essential behaviour, we allow for system reconfiguration and the introduction of new components.

Definition: A *feature for a PEPA model* consists of:

- one or more parameterised components which describe the behaviour of the newly added feature possibly re-using existing components;
- optionally, some non-parameterised components which are used to structure the new feature (components may be re-used here also); and
- a new *system equation* which describes how the new system is built from the existing one and the components of the two kinds described above.

3 Semantics of the feature construct

A feature will utilize a parameterized component and it will also typically make use of simple components. Incorporating a feature into a system is done by instantiating the parameter of the feature by the existing system, optionally combined with new simple components.

In operation, a feature monitors its base system. Whenever the base system makes a transition it is checked against the triggers for the feature. If it is not a trigger the base system proceeds as before. Otherwise, the transition is replaced as indicated and the feature determines how to proceed.

3.1 Impose and treat

A useful separation of concerns in describing feature integration is the distinction between *imposing* new behaviour on the system and *treating* the existing behaviour in a new way. In a state-based modelling approach such as ours this divides into redirecting the system evolution into new states in the former case and substituting new activities for existing ones in the latter case. In either case, some activity is used as a trigger for the new feature. If the activity happens then the new feature comes into effect. If not then the system behaves as before. Assume that the parameter P has an (α, r) transition to P', then there are four possible outcomes.

Treat (1):	$S(P) \xrightarrow{(\alpha,r)} S(P')$	α is not a trigger
Treat (2):	$S(P) \xrightarrow{(\alpha,r)} S'(P',Q)$	α is a trigger for Q
Impose (1):	$S(P) \xrightarrow{(\alpha,r)} S(P')$	α is not a trigger
Impose (2):	$S(P) \xrightarrow{(\beta,s)} S'(P',Q)$	α is a trigger for Q

Formally, in both cases there are two possibilities. Either the transition which the existing system would perform has a matching transition in the transitions of the parameterised component or it does not. In the cases where there is a matching transition the new state of the system is dictated by a parameterised component whose behaviour can depend on either P' or Q. In the cases where only one of the possible outcomes is of importance the parameterised component need only receive a single component as its parameter. In this case it might be that the derivative S' of S is itself. The Treat rule can introduce new behaviour. Additionally, the Impose rule is to replace an activity α performed at rate r by an activity β performed at rate s.

It is easy to see that the expressiveness of the Treat rule subsumes that of the Impose rule. When the metavariables α and β denote the same PEPA activity and the metavariables r and s denote the same PEPA rate then the Treat rule expresses the same adaptation of the existing system behaviour

Impose

$$\frac{P \xrightarrow{(\alpha,r)} P' \quad S(P) \not\xRightarrow{(\alpha,r)}}{S(P) \xrightarrow{(\alpha,r)} S(P')} \qquad \frac{P \xrightarrow{(\alpha,r)} P' \quad S(P) \xRightarrow{(\alpha,r)} S'(P',Q)}{S(P) \xrightarrow{(\alpha,r)} S'(P',Q)}$$

Treat

$$\frac{P \xrightarrow{(\alpha,r)} P' \quad S(P) \not\xRightarrow{(\alpha,-)}}{S(P) \xrightarrow{(\alpha,r)} S(P')} \qquad \frac{P \xrightarrow{(\alpha,r)} P' \quad S(P) \xRightarrow{(\alpha,-)} (\beta,s).S'(P',Q)}{S(P) \xrightarrow{(\beta,s)} S'(P',Q)}$$

Choose

$$\overline{S(P,Q) \longrightarrow S'(P)} \qquad \overline{S(P,Q) \longrightarrow S'(Q)}$$

Fig. 1. Semantics of the feature construct

as the Impose rule. It is also easy to see that the Impose rule by itself is not sufficient because it is not possible to prevent the exisiting system model from performing its *first* activity. This is a limitation because we wish to be able to impose new behaviour on *any* state of the exisiting system, even the initial state. The Treat rule does not suffer from this limitation.

Since the Treat rule subsumes the Impose rule, and is applicable in more situations, why then do we keep the Impose rule at all? The reason is that in practice it gives the most convenient form of expression to new features. The more general rule would almost always be used to simulate Impose.

It could be that we wish to make only a slight amendment to the existing system and it should not be the case that the feature construct *forces* reconfiguration. A simple example would be renaming activities as in "whenever the base system does activity α, do β instead and then continue as the base system". In telephony the Call Forwarding feature does this. This is achieved by furnishing the residual of a parameterised component with the two possible outcomes of the system as its parameters. Branching the system to track the evolution of its two possible futures on every feature application is of course impractical so we require the parameterised residual at the next step to select one of the derivatives with the Choose rules. Where appropriate we abbreviate pairing and choosing to follow only the selected component.

We make the preceeding informal description fully formal in the operational definition of the feature construct presented in Figure 1. This is built structurally from the transition relation on PEPA components (written $\longrightarrow$) and the transitions of parameterised components (written $\Longrightarrow$, we will use the notation $\not\Longrightarrow$ to indicate the absence of such a transition). The definition of the transition relation for PEPA components is in Figure 9 in Appendix A. The definition of the transition relation for parameterised components is in Figure 2. It depends on the definition given in Figure 3 of the judgement relation for contexts (written $\vdash$).

Overriding

$$C \vdash (\alpha, r).S(E) \xrightarrow{(\alpha,r)} S(E)$$

$$C \vdash (\alpha, \_).(\beta, s).S(E) \xrightarrow{(\alpha,\_)} (\beta, s).S(E)$$

Option

$$\frac{C \vdash E_1 \xrightarrow{(\alpha,r)} S(E)}{C \vdash E_1 + E_2 \xrightarrow{(\alpha,r)} S(E)} \qquad \frac{C \vdash E_2 \xrightarrow{(\alpha,r)} S(E)}{C \vdash E_1 + E_2 \xrightarrow{(\alpha,r)} S(E)}$$

$$\frac{C \vdash E_1 \xrightarrow{(\alpha,\_)} (\beta, s).S(E)}{C \vdash E_1 + E_2 \xrightarrow{(\alpha,\_)} (\beta, s).S(E)} \qquad \frac{C \vdash E_2 \xrightarrow{(\alpha,\_)} (\beta, s).S(E)}{C \vdash E_1 + E_2 \xrightarrow{(\alpha,\_)} (\beta, s).S(E)}$$

Application

$$\frac{C \vdash E' \xrightarrow{(\alpha,r)} S'(F)}{S(P) \xrightarrow{(\alpha,r)} S'(Q)} \qquad (S(E) \stackrel{\text{def}}{=} E' \text{ and } C \vdash E = P, F = Q)$$

$$\frac{C \vdash E' \xrightarrow{(\alpha,\_)} (\beta, s).S'(F)}{S(P) \xrightarrow{(\alpha,\_)} (\beta, s).S'(Q)} \qquad (S(E) \stackrel{\text{def}}{=} E' \text{ and } C \vdash E = P, F = Q)$$

Fig. 2. Transition relation for parameterised definitions

3.2   Parameterised definitions

Parameterised components are used to express the intervention of the new
feature on the existing system, as defined in Figure 2. We reuse the prefix
notation to express the activity of overriding the state and activities of the
existing system and the subsequent evolution to a new parameterised system.
In this reuse the two types of features are distinguished syntactically by the
number of activities which prefix their parameterised residual. If there is just
one activity then this is an *impose* feature and the activity performed will
be mirrored by the replacement activity. If there are two activities then the
first is the trigger for the feature (which will be absorbed) and the second is
the replacement activity. The rate of the triggering activity is not significant
and is denoted by an underscore. We use the plus notation in parameterised
components to express the option of monitoring a range of activities, summing
over all of the possibilities. By considering the syntactic form of such a sum it
is simple to derive statically the set of significant activities which can trigger
the activation of a newly defined feature. Just as significant is the use of
this set of activity names to determine which activities *cannot* cause the
invocation of a feature.

In the definition of the transition relation in Figure 2 the metavariable E used in parametric definitions is sometimes used to stand for a single component and sometimes for a pair of components, allowing a single rule to encompass both cases. A similar device could be used to allow a single trigger for a feature to be replaced by a sequence of activities.

3.3   Contexts

We introduce definition contexts to explain the binding of actual component parameters to formal parameter identifiers. We allow components to be parameterised by other components but the parameterisation is first-order, that is we do not allow parameterised components to be passed as parameters.

More formally, parameters range over PEPA expressions extended with formal parameter identifiers in addition to the identifiers used for PEPA constants. To preclude syntactic ambiguity, we use the convention that if a constant appears in a formal parameter expression then it denotes the component bound to that constant name and it is not a re-use of that identifier with another meaning in the body of the parameterised component definition. Thus it is not possible to make a hole in the scope of a component definition by re-using the identifier of that component as the identifier of a formal parameter.

Given the above syntactic restriction on formal parameter identifiers we note that contexts cannot contain re-definitions of identifiers of constants. The notation $C, I = P$ therefore denotes a context C extended by the binding of an identifier I to a component P. Such an expression can be used to judge that the identifiers I and P both denote the same component. The notation $E = P, L$ is used to denote a list of equalities beginning with one between E and P and continuing with those in L.

Note that, as is usual with definitional equality, the equality symbol used in contexts is not a commutative operator. In constrast the relations which are used to judge semantic equivalence between PEPA components such as PEPA's strong equivalence (also known as Markovian bisimulation) both preserve the familiar logical properties of equivalence relations and are congruences over all of the PEPA operators. This allows the component-wise substitution of equals for equals which is the foundation of a compositional modelling approach such as that embodied by PEPA.

$$\frac{}{C \vdash P = P} \qquad \frac{}{C, I = P \vdash I = P} \qquad \frac{C \vdash E = P \quad C \vdash L}{C \vdash E = P, L}$$

$$\frac{C \vdash E_1 = P_1 \quad C \vdash E_2 = P_2}{C \vdash E_1 \bowtie_L E_2 = P_1 \bowtie_L P_2} \qquad \frac{C \vdash E = P}{C \vdash E/L = P/L} \qquad \frac{C \vdash E = P}{C \vdash E = A}(A \stackrel{\text{def}}{=} P)$$

Fig. 3. The judgement relation for definition contexts

4 Example

We present here an example which serves only to help explain the use of the feature construct, not to act as a compelling defense of the use of formal notations in performance modelling. We consider first the very simple model of a transmitter which transmits at rate t to a receiver which receives data at rate r. This transmission is conducted through the medium of a network which passively cooperates with the transmitter and the receiver. This is the meaning of the $\top$ symbol used in the occurrences of the transmit and receive activities specified in the description of the network, that it is passive with respect to these activities.

The feature which we add to this simple system is a new component which monitors the network bandwidth which the transmitter has been able to obtain and signals whenever this drops below a critical threshold. In this circumstance the transmitter halves its transmission rate (for example, in a multimedia application by sampling an analogue input signal at half of the previous rate or in another application by applying data compression). The model is presented in Figure 4.

Existing system

$$Transmitter \stackrel{\text{def}}{=} (trans, t).Transmitter$$
$$Receiver \stackrel{\text{def}}{=} (recv, r).Receiver$$
$$Network \stackrel{\text{def}}{=} (trans, \top).Network'$$
$$Network' \stackrel{\text{def}}{=} (recv, \top).Network$$

$$System \stackrel{\text{def}}{=} Transmitter \underset{\{trans\}}{\bowtie} Network \underset{\{recv\}}{\bowtie} Receiver$$

Addition of monitoring feature

$$Transmitter' \stackrel{\text{def}}{=} (trans, t/2).Transmitter'$$
$$Monitor \stackrel{\text{def}}{=} (low, l).Monitor'$$
$$Monitor' \stackrel{\text{def}}{=} (high, h).Monitor$$

$$S(M \parallel (T \underset{\{trans\}}{\bowtie} N \underset{\{recv\}}{\bowtie} R)) \stackrel{\text{def}}{=}$$

$$(low, l).S(Monitor' \parallel (Transmitter' \underset{\{trans\}}{\bowtie} N \underset{\{recv\}}{\bowtie} R))$$
$$+$$
$$(high, h).S(Monitor \parallel (Transmitter \underset{\{trans\}}{\bowtie} N \underset{\{recv\}}{\bowtie} R))$$

$$System' \stackrel{\text{def}}{=} S(Monitor \parallel System)$$

Fig. 4. Use of the feature construct

It is possible that the congestion on the network will subsequently reduce. In this case the monitor will signal that it is suitable for the transmitter to resume transmitting at the higher rate. The details of how the monitor measures the consumption of the available bandwidth are abstracted away in the model and the signals to switch between transmitting at the low rate and the high rate are simply modelled by stochastic events with parameters l and h respectively. From a performance modelling point of view, the effect of adding the new feature is to turn the *Transmitter* component from a simple Poisson arrival process into a Markov modulated process. The new feature ensures that the correct *Transmitter* component is used whenever the *Monitor* component changes state. Upon witnessing a *low* signal from the monitor, the values held by the formal parameters M and T will be lost and the primed versions of the *Monitor* and *Transmitter* components will be used. Upon witnessing a *high* signal from the monitor, the values held by the formal parameters M and T will similarly be lost and the unprimed versions of the *Monitor* and *Transmitter* components will be reinstated.

The fact that there was no planning for the subsequent addition of a monitoring feature in the original model is reflected in the fact that the activities which are of interest to the monitoring feature are the *low* and *high* activities which are themselves newly added to the system via the *Monitor* component. We can see statically from the definition of the monitoring feature alone that the activities *trans* and *recv* are not used anywhere in the definition. This provides a simple and efficient check of freedom of interference between features and components.

5 Implementation

We have provided an embedding of the extended PEPA language in the higher-order functional programming language Standard ML [9] a language which we have previously used in the implementation of other software tools for the PEPA language, notably the PEPA Workbench [10].

The program structuring facilities of Standard ML are typical of those of a higher-order functional programming language. Functions are first-class and can be passed as parameters, returned as results, or used as values which can be stored in a data structure such as a list or a binary tree. The ability to manipulate functions in this way is crucially used in our embedding of PEPA components in Standard ML. In a language which did not support functions as first-class values the encoding of PEPA components would involve much more indirection. PEPA components describe the performance of activities by using conditional recursive definitions.

We map PEPA components onto Standard ML functions and provide datatype constructors which can be used to compose these functions as structured PEPA components can be built using the combinators of the language. Featured PEPA components are modelled as higher-order functions.

PEPA has four combinators, prefix (.), choice (+), co-operation (⋈, or ‖ if the co-operation set is empty) and hiding (/). These are mapped on to the constructors of a Standard ML datatype which can be used to compose component definitions. The constructors of this datatype are *, +, ⟨, ⟩, ‖ and /. These identifiers are chosen to resemble the corresponding symbols in the PEPA syntax, making their encoding relatively straightforward.

> **datatype** *Component* = * **of** (*Activity* * *Rate*) * (*unit* → *Component*)
> | + **of** *Component* * *Component*
> | ‖ **of** *Component* * *Component*
> | ⟨ **of** *Component* * *Activity* list
> | ⟩ **of** *Component* * *Component*
> | / **of** *Component* * *Activity* list

Fig. 5. Datatype definition for the PEPA combinators

To explain the use of these constructors, when the plus constructor is give a pair of components (*Component* * *Component*), it labels these as a choice, which is itself a kind of component.

The simple model from the previous section is encoded using our PEPA embedding in Standard ML in Figure 6. The datatype definitions at the

> **datatype** *Activity* = *trans* | *recv* | *tau* **of** {*hidden* : *Activity*};
> **datatype** *Rate* = *t* | *r* | *top*;
>
> **fun** *Transmitter* () = (*trans*, *t*) * *Transmitter*;
> **fun** *Receiver* () = (*recv*, *r*) * *Receiver*;
>
> **fun** *Network* () = (*trans*, *top*) * *Network*′
> **and** *Network*′ () = (*recv*, *top*) * *Network*;
>
> **fun** *System* () = *Transmitter* () ⟨[*trans*]⟩ *Network* () ⟨[*recv*]⟩ *Receiver* ();

Fig. 6. Implementation of the original system

beginning encode the names of activities and rates which are used in the model. The benefit which comes from this are the guarantees provided by the strict static type-checking of the Standard ML language. For example,

1. no activities or rates can be used within the model unless there is an accompanying declaration, and
2. an activity name cannot appear where a rate is expected.

The polymorphic type inference [11] mechanism of Standard ML means that this benefit is provided without the imposition of additional syntactic clutter such as typing assignments in function definitions.

The PEPA language defines a distinguished activity name, τ, which indicates that the activity is a private one. Components cannot co-operate on τ activities. The *tau* constructor of the activity datatype provides the ability to describe activity names such as *tau*{*hidden* = *recv*} which is a representation of a τ activity which additionally captures in the Standard ML encoding the information that the activity which was hidden was *recv*.

The PEPA language also defines a distinguished symbol $\top$ which is used to indicate passive co-operation in an activity. This is encoded as the constructor *top* in the datatype of activity rates.

Each of the PEPA component definitions translates into a function definition, introduced by the keyword 'fun'. Where a group of component definitions are mutually recursive—as *Network* and *Network'* are—they are introduced by a simultaneous binding where the function definitions are separated by the keyword 'and'. All functions in Standard ML have a single argument and so the definition of the function name is followed by Standard ML's *unit* pattern, analogous to the void type in Java. Each of the synchronisation sets used in the PEPA model contains only a single activity name but if more are required they can be included in lists such as [*trans*, *recv*].

For this simple model to be extended with the addition of the monitoring feature we need to extend the *Activity* datatype with the activity names *low* and *high* and to extend the *Rate* datatype with the rate *t_half*. We then add the new component definitions and the new feature, building the formal description of the new system from the description of the existing one by the application of the monitoring feature. The definitions of the transmitter and monitor components are straightforward and are omitted here. The accompanying definitions are presented in Figure 7. The monitoring feature is parameterised by a component. Using Standard ML's layered pattern matching, identifiers are provided for this parameter as a whole (*Param*) and for its subcomponents (M, T, N, R). The *step* function unrolls the component definition by a single transition returning the (*act, rate*) pair and the one-step derivative, *Param'*. The activity type determines the subsequent behaviour.

```
fun S (Param as (M || ((T ⟨[trans]⟩ N) ⟨[recv]⟩ R))) () =
  let val ((act, rate), Param') = step (Param) in
  case (act, rate) of
    (low, l) =>
      (low, l) * S(Monitor'() || ((Transmitter'() ⟨[trans]⟩ N) ⟨[recv]⟩ R))
  | (high, h) =>
      (high, h) * S(Monitor() || ((Transmitter() ⟨[trans]⟩ N) ⟨[recv]⟩ R))
  | default => (act, rate) * S(Param'())
  end;

val System' = S(Monitor() || System());
```

Fig. 7. Implementation of the monitoring feature

6 Related and future work

It is a good practice in any model development where a series of unpredictable revisions can occur to formally document system changes. A feature construct provides a formalism for such documentation. However, in the case of models which are used for performance analysis purposes there is an added urgency to formally document changes. The performance modelling process proceeds by specifying and evaluating measures of interest such as throughput, utilisation and bandwidth. These auxiliary definitions which accompany a system model are called *reward specifications*. Correct reward specifications are themselves difficult to obtain but the difficulty of relating reward specifications is acknowledged as being considerably more difficult again [12]. With a formal record of the changes to a model we at least have the promise of being able to automatically update the accompanying reward specifications. This remains as future work.

We are developing a model checker to allow formulae of the probabilistic modal logic PML_μ [13] to be checked over PEPA models which use the feature construct described here. Our present working implementation of the model checker has proven useful in detecting errors in PEPA models. The model checker has the useful facility to return a counterexample to show how the formula fails to be satisfied. This provides valuable guidance in the process of disgnosis of the critical flaw in the model and the subsequent re-working of the model to eliminate the error.

We have presented a method of describing additional features of a system separately from the description of the system itself. As features are progressively added, the complexity of a system inevitably grows. One avenue of future work is the investigation of the estimation of the additional complexity which is brought to the system by the addition of a single feature. Concretely, we could ask for a formula which computes how the size of the state space of the extended system will increase as a function of the state spaces of the systems which are used as actual parameters of parameterised components.

Conclusions

If performance modelling notations and tools are to realise the valuable contribution which they promise for the development of reliable and efficient complex software systems they must provide support not only for the initial design of systems but also for their correction, revision, and subsequent extension. We have shown how a new construct could be added to the PEPA performance modelling language: *parameterised components*. The new construct satisfies many of the desired goals of a feature construct and in addition promotes *structured whole-lifecycle performance modelling* of complex software systems. We have given the construct an operational semantics which builds upon the existing semantics of PEPA. We have applied the construct in a small case study.

Acknowledgements

The authors are grateful to Mark Ryan for suggestions which helped to improve the structure of the paper and the presentation of the feature construct in particular. This paper has benefited significantly from the helpful comments from the anonymous referees.

Stephen Gilmore is supported by Esprit Working group FIREworks and by the 'Distributed Commit Protocols' grant from the EPSRC. Jane Hillston is supported by the EPSRC 'COMPA' grant.

References

1. J. Hillston. *A Compositional Approach to Performance Modelling.* Cambridge University Press, 1996.
2. D.R.W. Holton. A PEPA specification of an industrial production cell. In S. Gilmore and J. Hillston, editors, *Proceedings of the Third International Workshop on Process Algebras and Performance Modelling*, pages 542–551. Special Issue of *The Computer Journal*, 38(7), December 1995.
3. S. Gilmore, J. Hillston, D.R.W. Holton, and M. Rettelbach. Specifications in Stochastic Process Algebra for a Robot Control Problem. *International Journal of Production Research*, 34(4):1065–1080, 1996.
4. A. El-Rayes, M. Kwiatkowska, and S. Minton. Analysing performance of lift systems in PEPA. In R. Pooley and J. Hillston, editors, *Proceedings of the Twelfth UK Performance Engineering Workshop*, pages 83–100, Department of Computer Science, The University of Edinburgh, September 1996.
5. H. Bowman, J. Bryans, and J. Derrick. Analysis of a multimedia stream using stochastic process algebra. In C. Priami, editor, *Sixth International Workshop on Process Algebras and Performance Modelling*, pages 51–69, Nice, September 1998.
6. L. Kloul, J.M. Fourneau, and F. Valois. Performance modelling of hierarchical cellular networks using PEPA. In J. Hillston, editor, *Proceedings of the Seventh International Workshop on Process Algebras and Performance Modelling*, Zaragosa, Spain, September 1999.
7. M. C. Plath and M. D. Ryan. Plug and play features. In W. Bouma, editor, *Feature Interactions in Telecommunications Systems V.* IOS Press, 1998.
8. M. D. Ryan. Feature-oriented programming: A case study using the SMV language. Technical report, School of Computer Science, The University of Birmingham, UK, September 1997.
9. R. Milner, M. Tofte, R. Harper, and D. MacQueen. *The Definition of Standard ML.* The MIT Press, 1996.
10. S. Gilmore and J. Hillston. The PEPA Workbench: A Tool to Support a Process Algebra-based Approach to Performance Modelling. In *Proceedings of the Seventh International Conference on Modelling Techniques and Tools for Computer Performance Evaluation*, number 794 in Lecture Notes in Computer Science, pages 353–368, Vienna, May 1994. Springer-Verlag.
11. R. Milner. A theory of type polymorphism in programming languages. *Journal of Computer and System Science*, 17(3):348–375, 1978.

12. J. Bradley and N. Thomas. Constructing a partial order for performance measures. In *Proceedings of the Sixteenth Annual UK Performance Engineering Workshop*, pages 177–186, Durham, United Kingdom, July 2000. UK Performance Engineering Workshop Press.
13. G. Clark, S. Gilmore, J. Hillston, and M. Ribaudo. Exploiting modal logic to express performance measures. In B.R. Haverkort, H.C. Bohnenkamp, and C.U. Smith, editors, *Computer Performance Evaluation: Modelling Techniques and Tools, Proceedings of the 11th International Conference*, number 1786 in LNCS, pages 211–227, Schaumburg, Illinois, USA, March 2000. Springer-Verlag.

A   Summary of the PEPA language

The PEPA language provides a small set of combinators. These allow language terms to be constructed defining the behaviour of components, via the activities they undertake and the interactions between them. The syntax may be formally introduced by means of the grammar shown in Figure 8.

$$S ::= \qquad \text{(sequential components)}$$
$$(\alpha, r).S \qquad \text{(prefix)}$$
$$| \quad S + S \qquad \text{(choice)}$$
$$| \quad C_S \qquad \text{(constant)}$$

$$P ::= \qquad \text{(model components)}$$
$$P \underset{L}{\bowtie} P \qquad \text{(cooperation)}$$
$$| \quad P/L \qquad \text{(hiding)}$$
$$| \quad C \qquad \text{(constant)}$$

Fig. 8. The syntax of PEPA

In the grammar S denotes a *sequential component* and P denotes a *model component* which executes in parallel. C stands for a constant which denotes either a sequential or a model component, as defined by a defining equation. C when subscripted with an S stands for constants which denote sequential components. The component combinators, together with their names and interpretations, are presented informally below.

Prefix: The basic mechanism for describing the behaviour of a system is to give a component a designated first action using the prefix combinator, denoted by a full stop. For example, the component $(\alpha, r).S$ carries out activity (α, r), which has action type α and an exponentially distributed duration with parameter r, and it subsequently behaves as S. Sequences of actions can be combined to build up a life cycle for a component.

Choice: The life cycle of a sequential component may be more complex than any behaviour which can be expressed using the prefix combinator alone. The choice combinator captures the possibility of competition between different possible activities. The component $P + Q$ represents a system which may behave either as P or as Q. The activities of both P and Q are enabled. The first activity to complete distinguishes one of them: the other is discarded. The system will behave as the derivative resulting from the evolution of the chosen component.

Constant: It is convenient to be able to assign names to patterns of behaviour associated with components. Constants are components whose meaning is given by a defining equation.

Hiding: The possibility to abstract away some aspects of a component's behaviour is provided by the hiding operator, denoted by the division sign in P/L. Here, the set L of visible action types identifies those activities which are to be considered internal or private to the component. These activities are not visible to an external observer, nor are they accessible to other components for cooperation. Once an activity is hidden it only appears as the unknown type τ; the rate of the activity, however, remains unaffected.

Cooperation: Most systems are comprised of several components which interact. In PEPA direct interaction, or *cooperation*, between components is represented by the butterfly combinator. The set which is used as the subscript to the cooperation symbol determines those activities on which the *cooperands* are forced to synchronise. Thus the cooperation combinator is in fact an indexed family of combinators, one for each possible *cooperation set* L (we write $P \parallel Q$ as an abbreviation for $P \bowtie_L Q$ when L is empty). When cooperation is not imposed, namely for action types not in L, the components proceed independently and concurrently with their enabled activities. However if a component enables an activity whose action type is in the cooperation set it will not be able to proceed with that activity until the other component also enables an activity of that type. The two components then proceed together to complete the *shared activity*. The rate of the shared activity may be altered to reflect the work carried out by both components to complete the activity.

In some cases, when an activity is known to be carried out in cooperation with another component, a component may be *passive* with respect to that activity. This means that the rate of the activity is left unspecified and is determined upon cooperation, by the rate of the activity in the other component. All passive actions must be synchronised in the final model.

Model components capture the structure of the system in terms of its *static* components. The dynamic behaviour of the system is represented by the evolution of these components, either individually or in cooperation. The form of this evolution is governed by a set of formal rules which give an operational semantics of PEPA terms. The semantic rules, in the structured operational style, are presented in Figure 9 without further comment; the interested reader is referred to [1] for more details. The rules are read as follows: if the transition(s) above the inference line can be inferred, then we can infer the transition below the line. The notation $r_\alpha(E)$ which is used in the third cooperation rule denotes the apparent rate of α in E.

Thus, as in classical process algebra, the semantics of each term in PEPA is given via a labelled *multi-transition* system—the multiplicities of arcs are significant. In the transition system a state corresponds to each syntactic term of the language, or *derivative*, and an arc represents the activity which causes one derivative to evolve into another. The complete set of reachable

Prefix

$$\frac{}{(\alpha, r).E \xrightarrow{(\alpha,r)} E}$$

Cooperation

$$\frac{E \xrightarrow{(\alpha,r)} E'}{E \bowtie_L F \xrightarrow{(\alpha,r)} E' \bowtie_L F} \ (\alpha \notin L) \qquad \frac{F \xrightarrow{(\alpha,r)} F'}{E \bowtie_L F \xrightarrow{(\alpha,r)} E \bowtie_L F'} \ (\alpha \notin L)$$

$$\frac{E \xrightarrow{(\alpha,r_1)} E' \quad F \xrightarrow{(\alpha,r_2)} F'}{E \bowtie_L F \xrightarrow{(\alpha,R)} E' \bowtie_L F'} \ (\alpha \in L) \text{ where } R = \frac{r_1}{r_\alpha(E)} \frac{r_2}{r_\alpha(F)} \min(r_\alpha(E), r_\alpha(F))$$

Choice

$$\frac{E \xrightarrow{(\alpha,r)} E'}{E + F \xrightarrow{(\alpha,r)} E'} \qquad\qquad \frac{F \xrightarrow{(\alpha,r)} F'}{E + F \xrightarrow{(\alpha,r)} F'}$$

Hiding

$$\frac{E \xrightarrow{(\alpha,r)} E'}{E/L \xrightarrow{(\alpha,r)} E'/L} \ (\alpha \notin L) \qquad\qquad \frac{E \xrightarrow{(\alpha,r)} E'}{E/L \xrightarrow{(\tau,r)} E'/L} \ (\alpha \in L)$$

Constant

$$\frac{E \xrightarrow{(\alpha,r)} E'}{A \xrightarrow{(\alpha,r)} E'} \ (A \stackrel{\text{def}}{=} E)$$

Fig. 9. The operational semantics of PEPA

states is termed the *derivative set* of a model and these form the nodes of the *derivation graph* formed by applying the semantic rules exhaustively.

The timing aspects of components' behaviour are not represented in the states of the derivation graph, but on each arc as the parameter of the negative exponential distribution governing the duration of the corresponding activity. The interpretation is as follows: when enabled an activity $a = (\alpha, r)$ will delay for a period sampled from the negative exponential distribution with parameter r. If several activities are enabled concurrently, either in competition or independently, we assume that a *race condition* exists between them. Thus the activity whose delay before completion is the least will be the one to succeed. The evolution of the model will determine whether

$$r_\alpha((\beta,r).P) = \begin{cases} r, & \alpha = \beta \\ 0, & \alpha \neq \beta \end{cases} \qquad r_\alpha(P + Q) = r_\alpha(P) + r_\alpha(Q)$$

$$r_\alpha(P/L) = \begin{cases} r_\alpha(P), & \alpha \notin L \\ 0, & \alpha \in L \end{cases} \quad r_\alpha(P \bowtie_L Q) = \begin{cases} r_\alpha(P) + r_\alpha(Q), & \alpha \notin L \\ \min(r_\alpha(P), r_\alpha(Q)), & \alpha \in L \end{cases}$$

Fig. 10. The apparent rate of α in PEPA components

the other activities have been *aborted* or simply *interrupted* by the state change. In either case the memoryless property of the negative exponential distribution eliminates the need to record the previous execution time.

When two components carry out an activity in cooperation the rate of the shared activity will reflect the working capacity of the slower component. We assume that each component has a capacity for performing an activity type α, which cannot be enhanced by working in cooperation (it still must carry out its own work), unless the component is passive with respect to that activity type. For a component P and an action type α, this capacity is termed the *apparent rate* of α in P (see Figure 10). It is the sum of the rates of the α type activities enabled in P. The apparent rate of α in a cooperation between P and Q over α will be the minimum of the apparent rate of α in P and the apparent rate of α in Q.

The derivation graph is the basis of the underlying Continuous Time Markov Chain (CTMC) which is used to derive performance measures from a PEPA model. The graph is systematically reduced to a form where it can be treated as the state transition diagram of the underlying CTMC. Each derivative is then a state in the CTMC. The *transition rate* between two derivatives P and Q in the derivation graph is the rate at which the system changes from behaving as component P to behaving as Q. It is denoted by $q(P,Q)$ and is the sum of the activity rates labelling arcs connecting node P to node Q. In order for the CTMC to be *ergodic* its derivation graph must be strongly connected. Some necessary conditions for ergodicity, at the syntactic level of a PEPA model, have been defined [1]. These syntactic conditions are imposed by the grammar introduced in Figure 8.

A.1   Availability of the modelling tools

The PEPA modelling tools, together with user documentation and papers and example PEPA models are available from the PEPA Web page at the address http://www.dcs.ed.ac.uk/pepa.

A Heuristic Algorithm to Detect Feature Interactions in Requirements

Maritta Heisel[1] and Jeanine Souquières[2]

[1] Fakultät für Informatik, Universität Magdeburg, D-39016 Magdeburg
[2] LORIA—Université Nancy2, B.P. 239 Bâtiment LORIA, F-54506
Vandœuvre-les-Nancy

Abstract. We present a heuristic algorithm to systematically detect feature interactions in requirements, which are expressed as constraints on system event traces.
The algorithm is part of a broader methodology for requirements elicitation and
formal specification. Given a new constraint and a set of already accepted constraints, it computes a set of candidate constraints that possibly interact with the
new one. We illustrate the algorithm by adding new features to a simple lift.

1 Introduction

The term "feature" has been coined in telecommunications, where a feature is
some service a client may subscribe to, such as call forwarding. It was also in
telecommunications where the problem of feature interaction occurred first.

Nowadays, features and the problem of integrating them are no longer
confined to the area of telecommunications. Features can be identified in
almost every software system, and there are even proposals to base the whole
software engineering process on features [3].

Following Turner et al. [3], we consider a feature as "a coherent and
identifiable bundle of system functionality that helps characterise the system
from the user perspective." More technically, a feature is "a clustering or
modularisation of individual requirements" of a requirements specification.
This definition emphasises the user-oriented nature of features.

When a system is described in terms of several features, the situation may
occur where the features make perfect sense when considered in isolation, but
their combination leads to contradictions or unwanted or unexpected system
behaviour. This situation is called *feature interaction.*

Although some authors (e.g., [4,5]) distinguish between "good" and "bad"
interactions, we prefer not to do so, because the question whether an interaction is desirable or not must be decided by the user and must only be
addressed when the *integration* of the various features is undertaken.

Different approaches to feature-oriented software development can be distinguished according to the following criteria:

1. How are features represented?
2. How are feature interactions detected/avoided?
3. How are feature interactions resolved?
4. How are features composed/integrated?

In our work, these questions are answered as follows:

1. Features are sets of formalised requirements, where each requirement is represented as a formula.
2. Because requirements are elicited in collaboration with the users by a brainstorming process, we do not attempt to avoid interactions right from the beginning but to detect and resolve them as early as possible. How we achieve this goal is the topic of this paper.
3. We do not attempt to resolve feature interactions automatically, because we think that each interaction should be discussed with the clients. It is then up to the clients to decide on the required system behaviour.
4. Feature composition corresponds to conjunction of formulae. Hence, it is commutative and associative. We think that these properties are desirable, because, for example, the behaviour of a telephone system should not depend on the order the user subscribed to the different features.

In this paper—which is a revised and extended version of the position paper [6]—we present an algorithm that, given a set of already accepted requirements and a new requirement to be added, calculates a set of candidate requirements with whom there might be an interaction. The algorithm is *heuristic*, which means that we cannot guarantee that all existing interactions are indeed detected.[1] It was developed as part of a method for requirements engineering, but it is also useful for the evolution of systems.

Section 2 gives a brief overview of our requirements elicitation method. Section 3 describes how to incorporate a single constraint into a set of existing constraints. Section 4 presents the algorithm for calculating interaction candidates. Section 5 illustrates the application of the approach by way of adding new features to a simple lift system. Related work is discussed in Section 6. Section 7 concludes the paper with a discussion of the approach and its benefits.

2   Method for Requirements Elicitation

Our method for requirements engineering [7,8] begins with an explicit requirements elicitation phase. The result of this first phase is a set of requirements, which are expressed formally as constraints on sequences of events or operations that can happen or be invoked in the context of the system. These constraints form the starting point for the development of a formal specification. In the present paper, however, we will not describe the specification phase, because the detection of feature interactions is part of the requirements elicitation phase.

Our approach to requirements engineering is inspired by the work of Jackson and Zave [9,10] and by the first steps of object oriented methods and

---

[1] Striving for a provably correct and complete algorithm would necessitate a formal and decidable notion of interaction. Because the notion of interaction covers more phenomena than just logical inconsistency, it is questionable if such a definition is possible or even desirable.

notations such as UML [11]. The starting point is a brainstorming process where the application domain and the requirements are described in natural language. This informal description is then transformed into a formal representation. Requirements elicitation is performed in five steps:

1. Introduce the domain vocabulary.
   The different notions of the application domain are expressed in a textual or graphical form.
2. State the facts, assumptions, and requirements concerning the system in natural language, as a set of fragments corresponding to parts of scenarios of the system behaviour.
   It does not suffice to just state requirements for the system. Often, facts and assumptions must be introduced to make the requirements satisfiable. *Facts* express conditions that always hold in the application domain, regardless of the implementation of the software system. Other requirements cannot be enforced, because e.g., human users might violate regulations. These conditions are expressed as *assumptions*.
3. List all relevant events that can happen in connection with the system, and classify them.
   Events concern the reactive part of the system. For each event, it must be stated who is in control of the event (the software system or its environment) and who can observe it.
4. List the system operations that can be invoked by users.
   This step is concerned with the non-reactive part of the system to be described. For purely reactive systems, it can be empty.
5. Formalise the facts, assumptions, and requirements as constraints on the possible traces of system events.

Using constraints to talk about the behaviour of the system has the following advantages:

- It is possible to express *negative* requirements, i.e., to require that certain things do not happen.
- It is possible to give scenarios, i.e., examples of system behaviour.
- Giving constraints does not fix the system behaviour entirely. The specification is not restricted unnecessarily. Any specification that fulfils the constraints is admitted [7].

Steps 1 through 4 can be carried out in any order or in parallel, with repetitions and revisions. There are validation conditions associated with the different steps, supporting quality assurance of the resulting product, stating necessary semantic conditions that the developed artifact must fulfil in order to serve its purpose properly:

- The vocabulary must contain exactly the notions occurring in the facts, assumptions, requirements, operations, and events.
- There must not be any events controlled by the software system and not shared with the environment.

3   Method to Incorporate Single Constraints

In Step 5 of the method, facts, assumptions, and requirements must be formalised one by one. But before a new formalised constraint is added to the set of already accepted constraints, its possible interactions with them should be analysed, in order to detect inconsistencies or undesired behaviour.

In the following, we will use the term *literal* to mean predicate or event symbols, or negations of such symbols. An event symbol e is supposed to mean "event e occurs", whereas $\neg e$ is supposed to mean "event e does not occur". If we refer to predicate symbols and their negations, we will use the term *predicate literal*. *Event literals* are defined analogously.

The following method gives guidelines how to incorporate a new constraint into a set of already existing constraints.

5.1   Formalise the new constraint as a formula on system traces.

To formalise facts, assumptions and requirements, we use traces, i.e., sequences of events happening in a given state of the system at a given time. The system is started in state S_1. When event e_1 happens at time t_1, then the system enters the state S_2, and so forth:

$$S_1 \xrightarrow[t_1]{e_1} S_2 \xrightarrow[t_2]{e_2} \ldots S_n \xrightarrow[t_n]{e_n} S_{n+1} \ldots$$

Let Tr be the set of possible traces. A constraint is expressed as a formula restricting the set Tr. For a given trace $tr \in Tr$, $tr(i)$ denotes the i-th element of this trace, $tr(i).s$ the state of the i-th element, $tr(i).e$ the event which occurs in that state, and $tr(i).t$ is the time at which the event occurs. For each possible trace, its prefixes are also possible traces. A formal specification of traces is given in Appendix A.

It may be necessary to introduce *predicates* on the system state to be able to express the constraints formally. For each predicate, events that establish it and events that falsify it must be stated. These events must be shared with the software system.

If possible, we recommend expressing constraints as implications, where either the precondition of the implication refers to an earlier state or an earlier point in time than the postcondition, or both the pre- and postcondition refer to the same state, i.e. we have an invariant of the system.

Example. When the lift is halted at a floor with the door open, a call for this floor is not taken into account.

$\forall tr : Tr;\ b : BUTTON \bullet \forall i : \operatorname{dom} tr \mid i \neq \#tr \bullet halted(tr(i).s)$
$\quad \wedge\ at(tr(i).s, floor(b)) \wedge door\_open(tr(i).s) \wedge tr(i).e = press(b)$
$\quad \Rightarrow \neg\ call(tr(i+1).s, floor(b))$

This formula contains the event symbol $press(b)$, which is parameterised with a button, and the predicate symbols *halted*, *at*, *door_open* and *call*. The predicates *at* and *call* have a floor as an additional parameter besides the system state. The function *floor* associates a floor with a button. The expression $\operatorname{dom} tr$ denotes the valid indices of the trace, i.e., $1 .. \#tr$, where $\#$ denotes the length of a trace.

5.2 Give a schematic expression of the constraint.

Our algorithm to determine interaction candidates uses schematic versions of formalised constraints. These have the form

$$x_1 \diamond x_2 \diamond \ldots \diamond x_n \rightsquigarrow y_1 \diamond y_2 \diamond \ldots \diamond y_k$$

where the x_i, y_j are literals and $\diamond$ denotes either conjunction or disjunction. The $\rightsquigarrow$ symbol separates the precondition from the postcondition. For transforming a constraint into its schematic form, we abstract from quantifiers and from parameters of predicate and event symbols.

Example. The schematic expression corresponding to the constraint stated before is

$$halted \wedge at \wedge door\_open \wedge press \rightsquigarrow \neg \ call$$

5.3 Update the tables of semantic relations.

Because our algorithm is completely automatic, it cannot be based on syntax alone. We also must take into account the semantic relations between the different symbols. We construct three tables of semantic relations:

1. Necessary conditions for events. If an event e can only occur if predicate literal pl is true, then this table has an entry $pl \leftarrow\!\!\sim e$.
   Example. The event *close* can only occur if the door is open:
   $door\_open \leftarrow\!\!\sim close$
2. Events establishing predicates. For each predicate literal pl, we need to know the events e that establish it: $e \rightsquigarrow pl$
   Example. The predicate *door_open* is established by the event *open*:
   $open \rightsquigarrow door\_open$
3. Relations between predicate literals. For each predicate symbol p, we determine:
   - the set of predicate literals it entails: $p_\Rightarrow = \{q : PLit \mid p \Rightarrow q\}$
   - the set of predicate literals its negation entails:
     $\neg \ p_\Rightarrow = \{q : PLit \mid \neg \ p \Rightarrow q\}$
   Example. $door\_open_\Rightarrow = \{\neg \ door\_closed, halted, at, \neg \ passes\_by\}$

These tables are not only useful to detect interactions; they are also useful to develop and validate the formal specification of the software system.

5.4 Determine interaction candidates, based on the list of schematic requirements (Step 5.2) and the semantic relation tables (Step 5.3).

The definition of the interaction candidates is given in Section 4.

5.5 Decide if there are interactions of the new constraint with the determined candidates.

The algorithm determines a set of candidates to examine. It does not prove that an interaction exists between the new constraint and each candidate. It is up to the analyst and the customer to decide if the conjunction of the new constraint with the candidates yields an unwanted behaviour or if it even is contradictory.

5.6 Resolve interactions.

To resolve an interaction, we usually relax requirements or strengthen assumptions. Once a constraint has been modified, an interaction analysis on those literals that were changed or newly introduced must be performed.

The following validation conditions are associated with Step 5 of the method for requirements elicitation:

- each requirement of Step 2 must be expressed,
- the set of constraints must be consistent,
- for each introduced predicate, events that modify it must be observable by the software system.

Steps 5.1 through 5.6 preserve the mutual coherence between the different constraints. Usually, revisions and communication with customers will be necessary.

4   Determining Interaction Candidates

Two constraints are interaction candidates for one another if they have overlapping preconditions but incompatible postconditions, as is illustrated in Figure 1. "Incompatible" does not necessarily mean "logically inconsistent"; it could also mean "inadequate" for the purpose of the system.

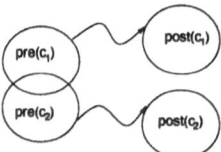

Fig. 1. Interaction candidates

Our algorithm to determine interaction candidates consists of two parts: precondition interaction analysis determines constraints with preconditions that are neither exclusive nor independent of each other. This means, there are situations where both constraints might apply. Their postconditions have to be checked for incompatibility. Postcondition interaction analysis, on the other hand, determines as candidates the constraints with incompatible postconditions. If in such a case the preconditions do not exclude each other, an interaction occurs.

4.1   Precondition Interaction Candidates

If two constraints[2] $\underline{x} \rightsquigarrow \underline{y}$ and $\underline{u} \rightsquigarrow \underline{w}$ have common literals in their precondition ($\underline{x} \cap \underline{u} \neq \varnothing$), then they are certainly interaction candidates.

---

[2] Underlined identifiers denote sets of literals.

But the common precondition may also be hidden. For example, if $\underline{x}$ contains the event e, $\underline{u}$ contains the predicate literal pl, and e is only possible if pl holds ($pl \leftsquigarrow e$), then we also have detected a common precondition pl of the two constraints.

The common precondition may also be detected via reasoning on predicates. If, for example, $\underline{x}$ contains the predicate literal pl, $\underline{u}$ contains the predicate literal q, and there is a predicate literal w with $pl \Rightarrow w$ and $q \Rightarrow w$, then w is a common precondition.

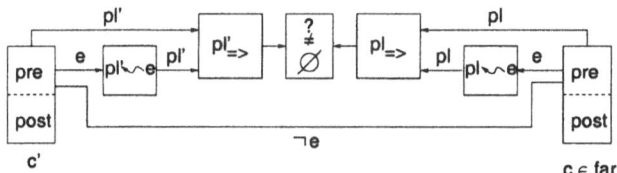

Fig. 2. Determining interaction candidates by precondition analysis

Figure 2 shows how to calculate interaction candidates $C_{pre}(c', far)$ by a precondition analysis for a new constraint c' with respect to the set *far* of facts, assumptions, and requirements already defined. Let

$$precond(x_1 \diamond x_2 \diamond \ldots \diamond x_n \leadsto y_1 \diamond y_2 \diamond \ldots \diamond y_k) = \{x_1, \ldots x_n\}$$

The set $pre\_predicates(c)$ of predicates that hold in the precondition of a constraint c are the predicate literals $pl \in precond(c)$ and the predicate literals pl with $pl \leftsquigarrow e$, for all event symbols $e \in precond(c)$:

$$\leftsquigarrow e = \{pl : PLit \mid pl \leftsquigarrow e\}$$
$$pre\_predicates(c) = (precond(c) \cap PLit) \cup \bigcup\nolimits_{e \in precond(c) \cap EVENT} \leftsquigarrow e$$

The *predicative closure* of the precondition of a constraint c results from the transitive and reflexive closure of the set $pre\_predicates(c)$ with respect to implication, i.e.

$$\bigcup\nolimits_{pl \in pre\_predicates(c)} pl_{\Rightarrow}$$

A constraint $c \in far$ is an interaction candidate for a new constraint c' if their preconditions or their respective predicative closures contain common literals.

$C_{pre}(c', far) =$
$\{c : far \mid precond(c) \cap precond(c') \neq \varnothing\} \cup$
$\{c : far \mid \exists pl : pre\_predicates(c); \; pl' : pre\_predicates(c') \bullet pl_{\Rightarrow} \cap pl'_{\Rightarrow} \neq \varnothing\}$

Two cases must be distinguished, because the precondition of a constraint can contain event literals, whereas the predicative closure of the precondition only contains predicate literals.

From the definition of $C_{pre}(c', far)$, it follows that the set of candidates is independent of the order in which the constraints are added, provided that the same tables of semantic relations are used to compute $\leftsquigarrow e$ and $pl_{\Rightarrow}$. Moreover, the candidate function distributes over set union of the preconditions of constraints:

$$\forall\, c, c_1, c_2 : Constraint;\ cs : \mathbb{P}\ Constraint \bullet$$
$$c_2 \in C_{pre}(c_1, cs \cup \{c_2\}) \Leftrightarrow c_1 \in C_{pre}(c_2, cs \cup \{c_1\})$$
$$\wedge \quad precond(c) = precond(c_1) \cup precond(c_2)$$
$$\Rightarrow C_{pre}(c, cs) = C_{pre}(c_1, cs) \cup C_{pre}(c_2, cs)$$

When a constraint is changed by adding a new literal to its precondition, a new interaction analysis has to be performed only on this new literal.

4.2 Postcondition Interaction Candidates

To find conflicting postconditions, we compute the predicative closure of the postcondition of the new constraint c' and the one of each constraint $c \in far$ in much the same way as for the preconditions. For an event e contained in the postcondition of a constraint, all predicate literals pl with $e \rightsquigarrow pl$ belong to the set $post\_predicates(c)$:

$$postcond(x_1 \diamond x_2 \diamond \ldots \diamond x_n \rightsquigarrow y_1 \diamond y_2 \diamond \ldots \diamond y_k) = \{y_1, \ldots y_k\}$$
$$e_\rightsquigarrow = \{pl : PLit \mid e \rightsquigarrow pl\}$$
$$post\_predicates(c) = (postcond(c) \cap PLit) \cup \bigcup\nolimits_{e \in postcond(c) \cap EVENT} e_\rightsquigarrow$$

A constraint c is an interaction candidate for the new constraint c' if there exists a literal pl in its postcondition or in its predicative closure, the negation of which is in the postcondition of c' or in its predicative closure. Figure 3 illustrates the definition.

$$C_{post}(c', far) =$$
$$\{c : far \mid postcond(c)\ opposite\ postcond(c')\}\ \cup$$
$$\{c : far \mid \exists\, pl : post\_predicates(c);\ pl' : post\_predicates(c') \bullet$$
$$pl_\Rightarrow\ opposite\ pl'_\Rightarrow\}$$
$$ls_1\ opposite\ ls_2 \Leftrightarrow \exists\, pl : ls_1 \bullet \neg\, pl \in ls_2$$

where ls_1, ls_2 are sets of literals and $\neg\,\neg\, l = l$.

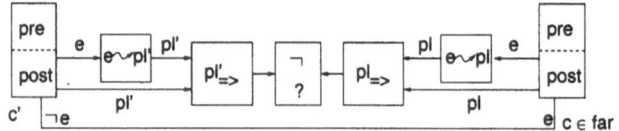

Fig. 3. Determining interaction candidates by postcondition analysis

Again, the two cases are necessary, because postconditions may contain event literals, whereas predicative closures only contain predicate literals.

Of course, this definition is symmetric, too, and C_{post} distributes over set union of postconditions of constraints.

The set of interaction candidates $C(c', far)$ of a new constraint c' with respect to the set far is the union of the precondition and the postcondition interaction candidates:

$$C(c', far) = C_{pre}(c', far) \cup C_{post}(c', far)$$

As already mentioned, computing the set $C(c', far)$ can be performed completely automatically. Moreover, the implementation is fairly simple and efficient. No theorem proving techniques or other search strategies are necessary.

5 Example: the Lift System

We first consider a simple lift with the following requirements:

1. The lift is called by pressing a button.
2. Pressing a call button is possible at any time.
3. A call is served when the lift arrives at the corresponding floor.
4. When the lift passes by a floor f, and there is a call from this floor, then the lift will stop at this floor.
5. When the lift has stopped, it will open the door.
6. When the lift door has been opened, it will close automatically after d time units.
7. The lift only changes its direction when there are no more calls in the current direction.
8. When the lift is halted at a floor with the door open, a call for this floor is not taken into account.
9. When the lift is halted at a floor with the door closed and receives a call for this floor, it opens its door.
10. Whenever the lift moves, its door must be closed.

As a fact, we formalise that the door can only be opened when it is closed and vice versa. Afterwards, we will add the following features:

11. When the lift is overloaded, the door will not close. Some passengers must get out.
12. The lift gives priority to calls from the executive landing.

5.1 Starting Point

Tables 1–4 present the schematic constraints for the fact and Requirements 1–10, and the corresponding tables of semantic relations. The formalised fact and Requirements 1–10 are given in Appendix B.

The schematic constraints, see Step 5.2 of the method of Section 3, are given in Table 1. In the formal expressions corresponding to $fact$, req_1 and req_7, the precondition refers to a later state than the postcondition, because necessary conditions for events to happen or predicates to be true are expressed. Our algorithm for feature interaction detection, however, requires the precondition to refer to an earlier or the same state as the postcondition. Hence, the schematic expressions for *fact*, *req₁* and *req₇* are based on the contraposition of the constraints given in Appendix B (i.e. $\neg Q \Rightarrow \neg P$ instead of $P \Rightarrow Q$).

The events establishing the predicates and their negations are given in Table 2, ordered alphabetically with respect to the predicate symbols. Table 3 shows the necessary conditions for the events, ordered alphabetically with respect to the event symbols. Finally, Table 4 shows the relations between the various predicate literals. This information is collected when performing Step 5.3 of the method of Section 3.

Constraint	schematic expression
fact	$\neg$ *door_closed* $\leadsto$ $\neg$ *open*
	$\neg$ *door_open* $\leadsto$ $\neg$ *close*
req₁	$\neg$ *press* $\leadsto$ $\neg$ *call*
req₂	*true* $\leadsto$ *press*
req₃	*at* $\leadsto$ $\neg$ *call*
req₄	*passes_by* $\wedge$ *call* $\leadsto$ *stop*
req₅	*stop* $\leadsto$ *open*
req₆	*open* $\leadsto$ *close*
req₇	*direction_up* $\wedge$ *call_from_up* $\leadsto$ *direction_up*
	direction_down $\wedge$ *call_from_down* $\leadsto$ *direction_down*
req₈	*halted* $\wedge$ *at* $\wedge$ *door_open* $\wedge$ *press* $\leadsto$ $\neg$ *call*
req₉	*halted* $\wedge$ *at* $\wedge$ *door_closed* $\wedge$ *press* $\leadsto$ *open*
req₁₀	$\neg$ *halted* $\leadsto$ *door_closed*

Table 1. Overview of schematic constraints

stop $\leadsto$ *at*	*close* $\leadsto$ *door_closed*
move $\leadsto$ $\neg$ *at*	*open* $\leadsto$ $\neg$ *door_closed*
press $\leadsto$ *call*	*open* $\leadsto$ *door_open*
stop $\leadsto$ $\neg$ *call*	*close* $\leadsto$ $\neg$ *door_open*
press $\leadsto$ *call_from_down*	*stop* $\leadsto$ *halted*
stop $\leadsto$ $\neg$ *call_from_down*	*move* $\leadsto$ $\neg$ *halted*
press $\leadsto$ *call_from_up*	*move* $\leadsto$ *passes_by*
stop $\leadsto$ $\neg$ *call_from_up*	*stop* $\leadsto$ $\neg$ *passes_by*

Table 2. Events establishing predicate literals

5.2 Adding new features

We now incorporate the features of overloading and executive floor, following the method of Section 3.

$$door\_open \;\leftsquigarrow\; close \qquad\qquad\qquad halted \;\leftsquigarrow\; move$$
$$at \;\leftsquigarrow\; move \qquad\qquad\qquad door\_closed \;\leftsquigarrow\; open$$
$$call \;\leftsquigarrow\; move \qquad\qquad\qquad \neg\, halted \;\leftsquigarrow\; stop$$
$$door\_closed \;\leftsquigarrow\; move \qquad\qquad\qquad passes\_by \;\leftsquigarrow\; stop$$

Table 3. Necessary conditions for events

$$at_\Rightarrow = \{halted, \neg\, passes\_by\}$$
$$\neg\, at_\Rightarrow = \{\neg\, halted, door\_closed, \neg\, door\_open, passes\_by\}$$
$$call_\Rightarrow = \varnothing$$
$$\neg\, call_\Rightarrow = \{\neg\, call\_from\_up, \neg\, call\_from\_down\}$$
$$call\_from\_down_\Rightarrow = \{call\}$$
$$\neg\, call\_from\_down_\Rightarrow = \varnothing$$
$$call\_from\_up_\Rightarrow = \{call\}$$
$$\neg\, call\_from\_up_\Rightarrow = \varnothing$$
$$door\_closed_\Rightarrow = \{\neg\, door\_open\}$$
$$\neg\, door\_closed_\Rightarrow = \{at, door\_open, halted, \neg\, passes\_by\}$$
$$door\_open_\Rightarrow = \{at, \neg\, door\_closed, halted, \neg\, passes\_by\}$$
$$\neg\, door\_open_\Rightarrow = \{door\_closed\}$$
$$halted_\Rightarrow = \{at, \neg\, passes\_by\}$$
$$\neg\, halted_\Rightarrow = \{\neg\, at, door\_closed, \neg\, door\_open, passes\_by\}$$
$$passes\_by_\Rightarrow = \{\neg\, at, door\_closed, \neg\, door\_open, \neg\, halted\}$$
$$\neg\, passes\_by_\Rightarrow = \{at, halted\}$$

Table 4. Relations between predicate literals

Requirement 11 (the *Overloaded* Feature). When the lift is overloaded, the door will not close. Some passengers must get out.

Step 5.1: Formalise the new constraint as a formula on system traces.
$$\forall\, tr : Tr \bullet \forall\, i : \mathsf{dom}\, tr \bullet overloaded(tr(i).s) \Rightarrow door\_open(tr(i).s)$$

Step 5.2: Give a schematic expression of the constraint.
$$overloaded \rightsquigarrow door\_open$$

Step 5.3: Update the tables of semantic relations.
With this constraint, we have introduced a new predicate symbol *overloaded* for which we must specify the events that modify it. Hence, we must introduce two new events *enter* and *leave*. We add the lines

$$enter \rightsquigarrow overloaded \qquad\qquad leave \rightsquigarrow \neg\, overloaded$$

to Table 2 and the lines

$$door\_open \leftsquigarrow enter \qquad\qquad door\_open \leftsquigarrow leave$$

to Table 3. To Table 4, we add the lines

$$overloaded_\Rightarrow = \{at, door\_open, \neg\, door\_closed, halted, \neg\, passes\_by\}$$
$$\neg\, overloaded_\Rightarrow = \varnothing$$

The entries of all predicates related to *overloaded* must be updated. We get the following changes:

$$\neg\, at_\Rightarrow = \{door\_closed, \neg\, door\_open, \neg\, halted, passes\_by,$$
$$\neg\, \mathbf{overloaded}\}$$
$$\neg\, door\_open_\Rightarrow = \{door\_closed, \neg\, \mathbf{overloaded}\}$$
$$door\_closed_\Rightarrow = \{\neg\, door\_open, \neg\, \mathbf{overloaded}\}$$
$$\neg\, halted_\Rightarrow = \{\neg\, at, door\_closed, \neg\, door\_open, passes\_by, \neg\, \mathbf{overloaded}\}$$
$$passes\_by_\Rightarrow = \{\neg\, at, door\_closed, \neg\, door\_open, \neg\, halted, \neg\, \mathbf{overloaded}\}$$

Step 5.4: Determine interaction candidates.
To determine the precondition interaction candidates, we determine the sets used in the definition of C_{pre} in Section 4.1:
$$pre\_predicates(req_{11}) = \{overloaded\}$$
Hence, the precondition interaction candidates are the ones that have one of the elements of $overloaded_\Rightarrow$ in their precondition, i.e., at, $door\_open$, $\neg\, door\_$ $closed$, $halted$, $\neg\, passes\_by$. According to Table 1, these are *fact* because of $\neg\, door\_closed$, req_3 because of at, req_8 because of at, $halted$ and $door\_open$, req_9 because of at and $halted$. Requirement req_2 is always a candidate for precondition interaction, because *true* is implied by every predicate.

To determine the postcondition interaction candidates, we proceed according to the definition of C_{post} in Section 4.2:
$$post\_predicates(req_{11}) = \{door\_open\}$$
Because $door\_open_\Rightarrow = \{at, \neg\, door\_closed, halted, \neg\, passes\_by\}$, we must look for postconditions that contain one of the elements of predicates $\neg\, at$, $door\_closed$, $\neg\, halted$, $passes\_by$ and related events that establish those predicates according to Table 2. These are *close* and *move*. According to Table 1, we get the candidates req_6 because of the event *close* and req_{10} because of $door\_closed$.

Step 5.5: Analyse possible interactions.
We do not have interactions with *fact*, req_2, req_3, req_8, req_9, req_{10}, but with req_6, because the door will not close automatically after d units time if the lift is overloaded.

Step 5.6: Eliminate interactions, if necessary. We relax req_6 as follows:

$$\forall\, tr : Tr \bullet \forall\, i : \mathsf{dom}\ tr \bullet tr(i).e = open \wedge last(tr).t > tr(i).t + d$$
$$\Rightarrow \exists\, j : \mathsf{dom}\ tr \bullet$$
$$(\mathbf{tr(j).t} \leq \mathbf{tr(i).t} + \mathbf{d} \wedge \mathbf{tr(j+1).t} > \mathbf{tr(i).t} + \mathbf{d} \wedge \neg\, \mathbf{overloaded(tr(j).s)})$$
$$\Rightarrow tr(j).e = close \wedge tr(j).t = tr(i).t + d)$$

The informal requirement req_6 has to be updated now. It becomes: "When the lift door has been opened, it will close automatically after d time units if the lift is not overloaded".

The new schematic constraint of req_6 becomes $open \rightsquigarrow close \vee overloaded$.
Since we have added the new postcondition *overloaded* to the constraint, we must now perform postcondition interaction analysis on this literal. With

$overloaded_\Rightarrow = \{at, door\_open, \neg\ door\_closed, halted, \neg\ passes\_by\}$ it follows
that we must look for constraints which contain one of the predicates $\neg\ at$,
$\neg\ door\_open$, $door\_closed$, $\neg\ halted$, $passes\_by$ in their postconditions and related events according to Table 2. These are *close* and *move*. In Table 1, we
find the candidate req_{10}. There is no interaction with it.

This concludes the introduction of the *overloaded* feature. To add this
feature to the lift, we not only had to introduce some new predicates and
change some requirements of the base system. More importantly, we had to
introduce two new events *enter* and *leave*. Our method requires that these
events be observable by the software system. Hence, a weight sensor must be
added to the lift if it is not already available.

Requirement 12 (the *Executive Floor* Feature). The lift gives priority
to calls from the executive landing.

Step 5.1: Formalise the new constraint as a formula on system traces.

$\forall tr : Tr \bullet \forall i : \mathsf{dom}\ tr \bullet$
    $call(tr(i).s, executive\_floor) \Rightarrow next\_stop(tr(i).s) = executive\_floor$

where $executive\_floor$ is a constant of type $FLOOR$.

Step 5.2: Give a schematic expression of the constraint.
    $call \rightsquigarrow next\_stop\_at\_executive\_floor$

Step 5.3: Update the tables of semantic relations.
With this constraint, we have introduced a new predicate symbol $next\_stop\_$
$at\_executive\_floor$ for which we must specify the events that modify it. We
add the lines
    $press \rightsquigarrow next\_stop\_at\_executive\_floor$
    $stop \rightsquigarrow \neg\ next\_stop\_at\_executive\_floor$
to Table 2. We add the following entry to Table 4:
    $next\_stop\_at\_executive\_floor_\Rightarrow = \{call\}$

Step 5.4: Determine interaction candidates.
To determine the precondition interaction candidates, we determine the sets
used in the definition of C_{pre} in Section 4.1:
$$pre\_predicates(req_{12}) = \{call\}$$
Hence, the precondition interaction candidates are the ones that have
one of the elements $call$, $call\_from\_up$, $call\_from\_down$ in their precondition.
According to Table 1, these are req_4 and req_7.

To determine the postcondition interaction candidates, we proceed according to the definition of C_{post} in Section 4.2:
$$post\_predicates(req_{12}) = \{next\_stop\_at\_executive\_floor\}$$
Because $next\_stop\_at\_executive\_floor_\Rightarrow = \{call\}$, we must look for postconditions that contain the predicate $\neg\ call$ and related events according to
Table 2, that is *stop*. According to Table 1, we get as candidates req_1, req_3
and req_8 because of $\neg\ call$ and req_4 because of the event *stop*.

Step 5.5: Analyse possible interactions.
We have no interactions with req_1, req_3 and req_8, but with req_4 and req_7, because req_{12} gives priority to the executive floor and not to the current floor as expressed in req_4 or to the current direction as expressed in req_7.

Step 5.6: Eliminate interactions, if necessary.
To adjust req_4, we add a new precondition to it; req_4 becomes

$$\forall\, tr : Tr;\ f : FLOOR \bullet (\textbf{let } tr' == remove(tr, \{b : Button \bullet press(b)\}) \bullet$$
$$\forall\, i : \text{dom } tr' \mid i \neq \#tr' \bullet passes\_by(tr'(i).s, f) \wedge call(tr'(i).s, f)$$
$$\wedge\ (\textbf{f = executive\_floor} \vee \neg\ \textbf{call(tr'(i).s, executive\_floor}))$$
$$\Rightarrow tr'(i+1).e = stop)$$

The informal requirement req_4 has to be updated. It becomes: "When the lift passes by a floor f, and there is a call from this floor, then the lift will stop at this floor if f is the executive floor or there is no call from the executive floor".

The new schematic expression for req_4 is:

$$passes\_by \wedge call \wedge (passes\_by\_executive\_floor \vee \neg\ call) \rightsquigarrow stop$$

Note that now we have *call* as well as $\neg$ *call* in the schematic precondition of the constraint. This is not a contradiction (*call* and $\neg$ *call* have different arguments), but only enlarges the set of possible interaction candidates.

Moreover, to capture the new precondition $f = executive\_floor$, we have introduced a new predicate symbol *passes_by_executive_floor* with

$$next\_stop\_at\_executive\_floor_{\Rightarrow} =$$
$$\{passes\_by, \neg\ at, door\_closed, \neg\ door\_open, \neg\ halted\}$$

We must now perform a precondition interaction analysis on the new preconditions $\neg$ *call* and *passes_by_executive_floor*. Concerning $\neg$ *call*, our candidates are the constraints with precondition $\neg$ *call*, $\neg$ *call_from_up*, $\neg$ *call_from_down*, because there are no related events. There are no interaction candidates. Concerning *passes_by_executive_floor*, we also do not get any new candidates, because all candidates were already candidates because of *passes_by*.

To adjust req_7, we also add new preconditions.

$$\forall\, tr : Tr \bullet \forall\, i : \text{dom } tr \mid i \neq \#tr \bullet$$
$$(direction(tr(i).s) = up \wedge direction(tr(i+1).s) = down$$
$$\Rightarrow (\neg\ call\_from\_up(tr(i).s) \vee \textbf{call(tr(i).s, executive\_floor})))$$
$$\wedge\quad (direction(tr(i).s) = down \wedge direction(tr(i+1).s) = up$$
$$\Rightarrow (\neg\ call\_from\_down(tr(i).s) \vee \textbf{call(tr(i).s, executive\_floor})))$$

The informal requirement req_7 has to be updated. It becomes: "The lift only changes its direction when there are no more calls in the current direction or there is a call from the executive floor".

The new schematic expressions for req_7 are:

$$direction\_up \wedge call\_from\_up \wedge \neg\ call \rightsquigarrow direction\_up$$
$$direction\_down \wedge call\_from\_down \wedge \neg\ call \rightsquigarrow direction\_down$$

As for req_4, we must perform a precondition interaction analysis on the new precondition $\neg\ call$. This yields the same candidates as before, plus the new version of req_4. Again, there is no further interaction.

6   Related work

In general, there are two ways to deal with the feature interaction problem. The first way is to *prevent* feature interactions right from the beginning, for example by enforcing modularity in the design of features. This approach is advocated by Jackson and Zave in their Distributed Feature Composition (DFC) virtual architecture [12]. Preventing feature interactions is supported by making feature first-class citizens in specification languages. For example, Plath and Ryan add a feature construct to the SMV language [13].

The second way to deal with feature interactions is to *detect* interactions and then resolve them. Even when the goal is to prevent feature interactions, algorithms for detecting them are indispensable. Zave [5] presents a method for preventing feature interaction problems. She points out that some interactions are desirable and that her method needs an analysis algorithm that generates a list of possible interactions among a set of features. Feature designers must adjust the feature specifications in an iterative process until the only remaining interactions are desirable ones.

How interactions can be detected depends on how features are specified and how interactions are defined. Bruns et al. [4] distinguish the following approaches:

- In the logical approach, features are specified as logical formulas and feature composition is logical conjunction. Feature interaction occurs if two features cannot be simultaneously satisfied.
- In the network specification approach [14], features are specified as sets of traces of network events and feature composition is set union. Feature interaction occurs if two feature sets intersect after certain operations are performed.
- In the operational approach [15], features are specified as processes and feature composition is some concurrent composition operation. Feature interaction occurs if the composed features fail to satisfy a global property such as deadlock freedom.
- In the feature as service transformer approach [4], a service describes the behaviour of a server that responds to input events. Feature composition is the successive transformation of a service by a sequence of features. Two notions of feature interaction are defined: two features interact (1) if the order in which they are applied affects the system behaviour, (2) if in some state, an input generates outputs that interfere.

Our work is a logical approach. What distinguishes it from other logical approaches, however, is the fact that we do not equate feature interaction with logical contradiction. In our opinion, logical contradiction is a sufficient but not a necessary condition for a feature interaction to appear. An example will be given in Section 7.

Other recent logical approaches are described in [13,16–18]. All of them use model checking techniques to detect interactions.

Jonsson et al. [16] propose a technique for hierarchically structuring requirements specifications in a way that simplifies change management and supports validation. As in our approach, requirements can be formulated and updated incrementally, supporting an evolutionary modelling of the application domain. Validation consists in checking iteratively the initial set of requirements (expressed in a linear time temporal logic) against the system model (expressed as a collection of automata), in scenarios where only a single feature is activated.

Like Jonsson et al., Felty and Namjoshi [17] use temporal logic to specify features and apply model checking to detect inconsistencies in the specification. In contrast to [16], they do not set up a separate system model, but convert feature specifications in ω-automata to perform the satisfiability test.

Khoumsi and Bevelo [18] have identified different kinds of interactions. Their interaction detection procedure is based on the search of special properties such as feature termination, variable consistency, events compatibility, event delayability and dependence on variables.

All of these detection procedures work on simplified models of the system. For example, telephone networks with only a fixed (and small) number of telephones are considered. The idea to detect interactions on simplified requirements or system models pertains also to our algorithm.

The work discussed so far is specifically designed to cope with the feature interaction problem. However, the notion of a *goal* [19] also provides a firm basis for detecting interactions between requirements. Van Lamsweerde and Letier use the concept of an *obstacle* or goal obstruction, which defines undesirable behaviour, to produce a refinement tree, the root of which is a goal negation. They define heuristics and formal techniques [20] to systematically generate obstacles from goal specifications and domain properties.

7   Discussion

We have presented an algorithm that helps to detect interactions in requirements. The algorithm is part of a more general method to systematically perform the first phases of software development. A systematic analysis of interactions leads to a better understanding of the requirements and avoids costly changes in later phases. The approach we have presented is domain independent and is method rather than language oriented.

It is useful not only for new systems but also for the evolution of systems. System evolution is motivated by changing requirements. Either new requirements are introduced or old requirements are replaced by different ones. In much the same way as for new systems, our algorithm can be used to analyse the consequences of changing the requirements before any changes to the software system are made.

We find it important that the detection of feature interactions be independent of the order in which the features are added, because this order may be arbitrary and insignificant. Moreover, we do not attempt to resolve feature interactions automatically. Such decisions are best taken by the customers.

We have already noted that it is important to find logical contradictions in requirements, but that not all interactions amount to logical contradictions. In the case study of an access control system [21], we had the following requirements: "when the door is unblocked, it will be re-blocked after 30 seconds" and "when a person has entered the building, the door will be re-blocked". These requirements interact, because it is intended to block the door immediately after the person has entered and not only after 30 seconds. Logically, however, the two requirements are not contradictory. It would suffice to re-block the door after 30 seconds, no matter if the person has entered or not. Hence, our algorithm can detect interactions that cannot be detected with logical procedures that detect only contradictions.

The approach for detecting feature interactions is truly heuristic. Its virtue lies in the fact that interactions on the requirements level can be detected very early, before a formal specification is set up, and with relatively little effort. Even though determining the interaction candidates is tedious if performed by hand, the procedures to determine the sets C_{pre} and C_{post} as defined in Section 4 are very easy to implement. Theorem proving techniques are unnecessary. Using our procedure, customers must inspect much fewer candidates than if a complete analysis, i.e. an inspection of all previously accepted constraints, were performed.

The semantic information collected in the tables of necessary conditions for events, events establishing predicate literals, and relations between predicate literals not only contributes to a better understanding of the requirements, but also greatly facilitates the process of setting up and validating a formal specification for the software system to be built, as is shown in [7,8].

Acknowledgement. We thank Thomas Santen for his comments on this paper.

References

1. M. Calder and E. Magill, editors. *Proc. 6th Feature Interaction Workshop, FIW 2000*. IOS Press Amsterdam, 2000.
2. K. Kimbler and W. Bouma, editors. *Proc. 5th Feature Interaction Workshop, FIW 1998*. IOS Press Amsterdam, 1998.
3. R. Turner, A. Fuggetta, L. Lavazza, and A. Wolf. A conceptual basis for feature engineering. *Journal of Systems and Software*, 49(1):3–15, 1999.
4. G.B. Bruns, P. Mataga, and I. Sutherland. Features as Service Transformers. In Kimbler and Bouma [2], pages 85–97.
5. P. Zave. Systematic design of call-coverage features. In *Proc. 7th International Conference on Algebraic Methodology and Software Technology*, LNCS 1548. Springer-Verlag, 1999.
6. M. Heisel and J. Souquières. A heuristic approach to detect feature interactions in requirements. In Kimbler and Bouma [2], pages 165–171.
7. M. Heisel and J. Souquières. A Method for Requirements Elicitation and Formal Specification. In J. Akoka and M. Bouzeghoub and I. Comyn-Wattiau and E. Métais, editor, *Proceedings of the 18th International Conference on Conceptual Modeling*, LNCS 1728, pages 309–324. Springer Verlag, November 1999.

8. M. Heisel and J. Souquières. De l'élicitation des besoins à la spécification formelle. *Technique et science informatiques*, 18(7):777–801, 1999.
9. M. Jackson and P. Zave. Deriving Specifications from Requirements: an Example. In *Proceedings 17th Int. Conf. on Software Engineering, Seattle, USA*, pages 15–24. ACM Press, 1995.
10. P. Zave and M. Jackson. Four dark corners of requirements engineering. *ACM Transactions on Software Engineering and Methodology*, 6(1):1–30, January 1997.
11. M. Fowler and K. Scott. *UML distilled. Applying the standard Object Modelling Language*. Addison-Wesley, 1997.
12. M. Jackson and P. Zave. Distributed Feature Composition: A Virtual Architecture for Telecommunications Services. *IEEE Transactions on Software Engineering*, 24(10):831–847, October 1998.
13. M. Plath and M. Ryan. Plug-and-Play features. In Kimbler and Bouma [2], pages 150–164.
14. A. Aho, S. Gallagher, N. Griffeth, C. Schell, and D. Swayne. Scf3/sculptor with Chisel: Requirements engineering for communication services. In Kimbler and Bouma [2], pages 45–63.
15. K.E. Cheng. Towards a Formal Model for Incremental Service Specification and Interaction Management Support. In L.G. Bouma and H. Velthuijsen, editors, *Feature Interaction in Telecommunication*. IOS Press Amsterdam, 1994.
16. B. Jonsson, T. Margaria, G. Naeser, J. Nystrom, and B. Steffen. Incremental Requirement Specification for Evovlving Systems. In Calder and Magill [1], pages 145–162.
17. A. Felty and K. Namjoshi. Feature specification and automatic conflict detection. In Calder and Magill [1], pages 179–192.
18. A. Khoumsi and R.J. Bevelo. A detection method developed after a thorough study of the contest held in 1998. In Calder and Magill [1], pages 226–240.
19. A. van Lamsweerde and E. Letier. Integrating Obstacles in Goal-directed Requirements. In *Proc. of the 20 th International Conference on Software Engineering, ICSE'98*, Kyoto, Japan, 1998. IEEE.
20. A. van Lamsweerde and E. Letier. Handling obstacles in goal-directed requirements engineering. *IEEE Transactions on Software Engineering*, 2000. Special Issue on Exception Handling.
21. J. Souquières and M. Heisel. Une méthode pour l'élicitation des besoins: application au système de contrôle d'accès. In Yves Ledru, editor, *Proceedings Approches Formelles dans l'Assistance au Développement de Logiciels - AFADL'2000*, pages 36–50. LSR-IMAG, Grenoble, 2000. http://www-lsr.imag.fr/afadl/Programme/ProgrammeAFADL2000.html.
22. J. M. Spivey. *The Z Notation – A Reference Manual*. Prentice Hall, 2nd edition, 1992.

A   Formal Expression of Constraints on Traces

In the following specification of system traces, we use the Z notation [22]. Each trace of the system is a sequence of trace items, where events later in the sequence must not happen earlier in time than events earlier in the sequence. The sign $\leq_t$

denotes a relation "not later" on time, which fulfils the axioms of a partial ordering relation.

$[STATE, EVENT, TIME]$

```
┌─ TraceItem ─────────────────────────────
│ s : STATE
│ e : EVENT
│ t : TIME
```

For each system, we will call the set of admissible traces Tr. Constraints will be expressed as formulas restricting the set Tr. For each possible trace, its prefixes are also possible traces.

```
┌ TRACE : ℙ(seq TraceItem)
├──────────────────────────────────
│ ∀ tr : TRACE • ∀ i : dom tr • i = #tr ∨
│     (tr i).t ≤_t (tr(i + 1)).t
```

```
┌ Tr : ℙ TRACE
├──────────────────────────────────
│ ∀ tr : Tr; tr' : TRACE |
│     tr' prefix tr • tr' ∈ Tr
```

The function *remove* takes a trace and a set of events as its arguments and removes all trace elements whose event is in the given set.

```
┌ remove : TRACE × ℙ EVENT → TRACE
├──────────────────────────────────
│ ∀ tr : TRACE; evs : ℙ EVENT •
│     remove(tr, evs) = tr ↾ {ti : TraceItem | ti.e ∉ evs}
```

B   Formal Versions of Requirements and Facts

Fact. The door can only be opened when it is closed and vice versa.

$\forall tr : Tr • \forall i : dom\ tr •$
$\quad (tr(i).e = open \Rightarrow door\_closed(tr(i).s) \wedge$
$\quad (tr(i).e = close \Rightarrow door\_open(tr(i).s)$

Requirement 1. The lift is called by pressing a button.

$\forall tr : Tr; b : BUTTON • \forall i : dom\ tr •$
$\quad call(tr(i).s, floor(b)) \Rightarrow (\exists j : dom\ tr \mid j < i • tr(j).e = press(b))$

Requirement 2. Pressing a call button is possible at any time.

$\forall tr : Tr; b : BUTTON • \exists tr' : Tr • front(tr') = tr \wedge last(tr').e = press(b)$

Requirement 3. A call is served when the lift arrives at the corresponding floor.

$\forall tr : Tr; f : FLOOR • \forall i\ dom\ tr • at(tr(i).s, f) \Rightarrow \neg\ call(tr(i).s, f)$

Requirement 4. When the lift passes by a floor f, and there is a call from this floor, then the lift will stop at this floor.

$$\forall\, tr : TR;\; f : FLOOR \bullet (\mathbf{let}\ tr' == remove(tr, \{b : BUTTON \bullet press(b)\}) \bullet$$
$$\forall\, i : \mathbf{dom}\ tr' \mid i \neq \#tr' \bullet$$
$$passes\_by(tr'(i).s, f) \wedge call(tr'(i).s, f) \Rightarrow tr'(i+1).e = stop)$$

Because *press* events are always possible, we must remove them from the traces (see Appendix A) when we want to express liveness conditions for the lift.

Requirement 5. When the lift has stopped, it will open the door.

$$\forall\, tr : Tr;\; f : FLOOR \bullet (\mathbf{let}\ tr' == remove(tr, \{b : BUTTON \bullet press(b)\}) \bullet$$
$$\forall\, i : \mathbf{dom}\ tr' \mid i \neq \#tr' \bullet tr'(i).e = stop \Rightarrow tr'(i+1).e = open)$$

Requirement 6. When the lift door has been opened, it will close automatically after d time units.

$$\forall\, tr : Tr \bullet \forall\, i : \mathbf{dom}\ tr \bullet tr(i).e = open \wedge last(tr).t > tr(i).t + d$$
$$\Rightarrow \exists\, j : \mathbf{dom}\ tr \bullet tr(j).e = close \wedge tr(j).t = tr(i).t + d$$

Requirement 7. The lift only changes its direction when there are no more calls in the current direction.

$$\forall\, tr : Tr \bullet \forall\, i : \mathbf{dom}\ tr \mid i \neq \#tr \bullet$$
$$(direction(tr(i).s) = up \wedge direction(tr(i+1).s) = down$$
$$\Rightarrow \neg\ call\_from\_up(tr(i).s))$$
$$\wedge \quad (direction(tr(i).s) = down \wedge direction(tr(i+1).s) = up$$
$$\Rightarrow \neg\ call\_from\_down(tr(i).s))$$

Requirement 8. When the lift is halted at a floor with the door open, a call for this floor is not taken into account.

$$\forall\, tr : Tr;\; b : BUTTON \bullet \forall\, i : \mathbf{dom}\ tr \mid i \neq \#tr \bullet halted(tr(i).s)$$
$$\wedge at(tr(i).s, floor(b)) \wedge door\_open(tr(i).s) \wedge tr(i).e = press(b)$$
$$\Rightarrow \neg\ call(tr(i+1).s, floor(b))$$

Requirement 9. When the lift is halted at a floor with the door closed and receives a call for this floor, it opens its door.

$$\forall\, tr : Tr;\; b : BUTTON \bullet \forall\, i \in \mathbf{dom}\ tr \bullet halted(tr(i).s)$$
$$\wedge at(tr(i).s, floor(b)) \wedge door\_closed(tr(i).s) \wedge tr(i).e = press(b)$$
$$\Rightarrow ((\exists\, j : \mathbf{dom}\ tr \bullet j > i \wedge \forall\, b : BUTTON \bullet tr(j) \neq press(b))$$
$$\Rightarrow (\exists\, k : \mathbf{dom}\ tr \mid k > i \bullet tr(k).e = open \wedge$$
$$\forall\, l \in i+1\,..\,k-1 \bullet \exists\, b : BUTTON \bullet tr(l).e = press(b)))$$

Requirement 10. Whenever the lift moves, its door must be closed.

$$\forall\, tr : Tr \bullet \forall\, i : \mathbf{dom}\ tr \bullet \neg\ halted(tr(i).s) \Rightarrow door\_closed(tr(i).s)$$

Defining Features for CSP: Reflections on the Feature Interaction Contest

Malte Plath[1] and Mark Dermot Ryan[1]

School of Computer Science, University of Birmingham, Edgbaston, Birmingham B15 2TT, England. mcp,mdr@cs.bham.ac.uk

1   Introduction

The second Feature Interaction Contest was held in conjunction with the 6th International Workshop on Feature Interactions in Telecommunication and Software Systems (FIW'00) [1]. The aim of the contest was to compare various methods and tools for detecting feature interactions. To enable a comparison, the contest's objective was to detect interactions among a given set of features for a given telephone system. The contest instructions contained detailed (albeit imprecise) specifications of the base system and twelve features.

In this paper we describe how we tackled the contest and the experiences we had in the process. In the following two sections we briefly describe how the contest was set out and what implications that had for detecting feature interactions. We then detail the methods we used and the results we obtained before summing up our experiences in section 7.

We used a combination of two techniques: static (syntactic) analysis and model checking using the model checker FDR [5]. While we had initially planned to integrate the features into the base system and then to detect interactions by model checking, it became clear very quickly that a simple syntactic analysis of the features was sufficient to detect a large proportion of the interactions. Hence, in this paper, we first describe this syntactic method (section 4), before showing how we used model checking to find more interactions (section 5). Finally, we give an overview of the full results of our analysis in section 5.4; a complete and detailed description of the results can be found in [4].

We are delighted to report that our entry won the contest in the category of two-feature interactions. For a summary of the results produced by each of the contestants, see [2].

2   The Contest Model

The base system for the contest is a model of the plain old telephone service (POTS) given as a labelled transition diagram (Figure 1) for a single telephone line (basic call model, BCM) and a description of the network architecture, a simple star network, with a single switch relaying information

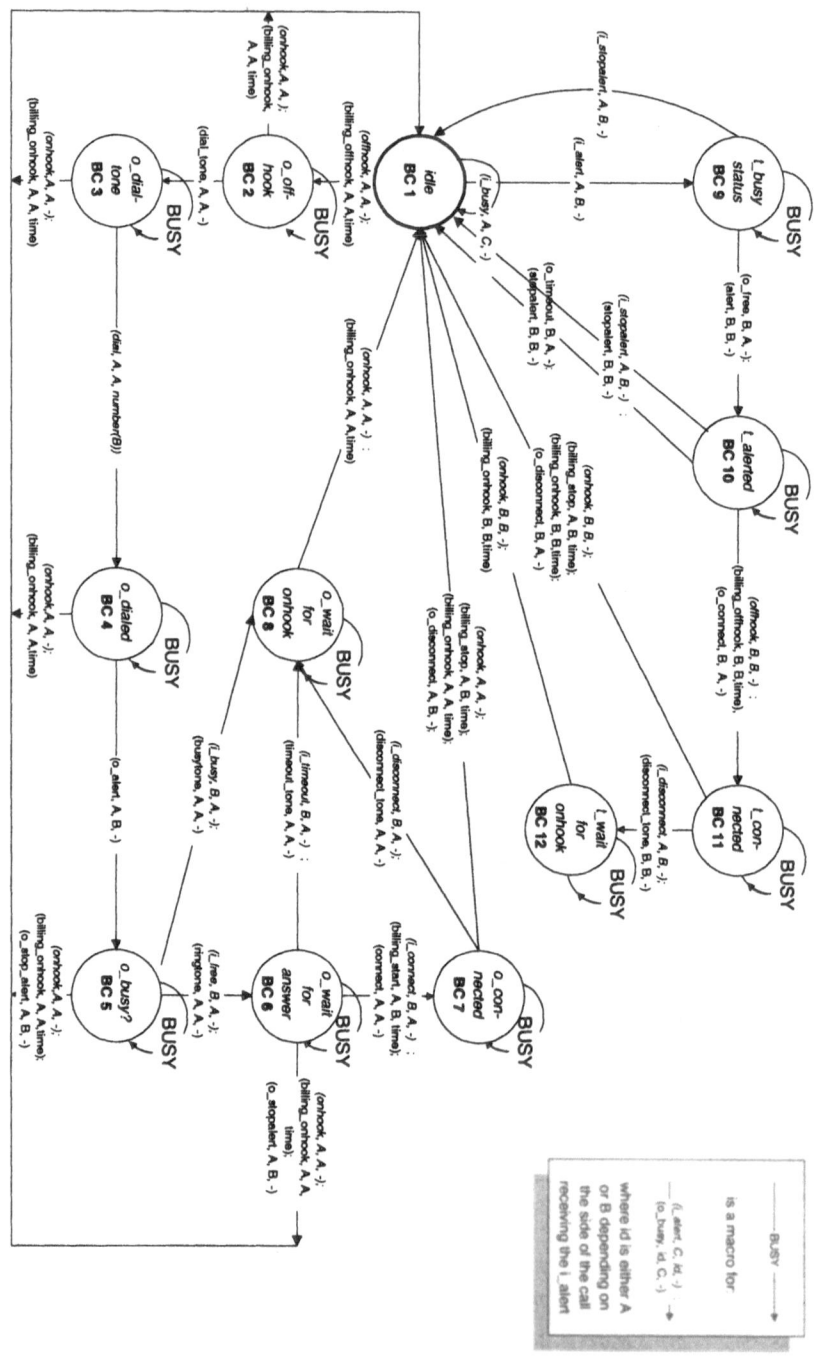

Fig. 1. The Basic Call Model [3]

between lines, and a billing database, of which no details are given in the contest instructions [3].

Most transitions are labelled with multiple messages, some of which denote local events, such as user input or signals to the user, others stand for messages sent to or received from other phones in the network; finally, there are billing messages.

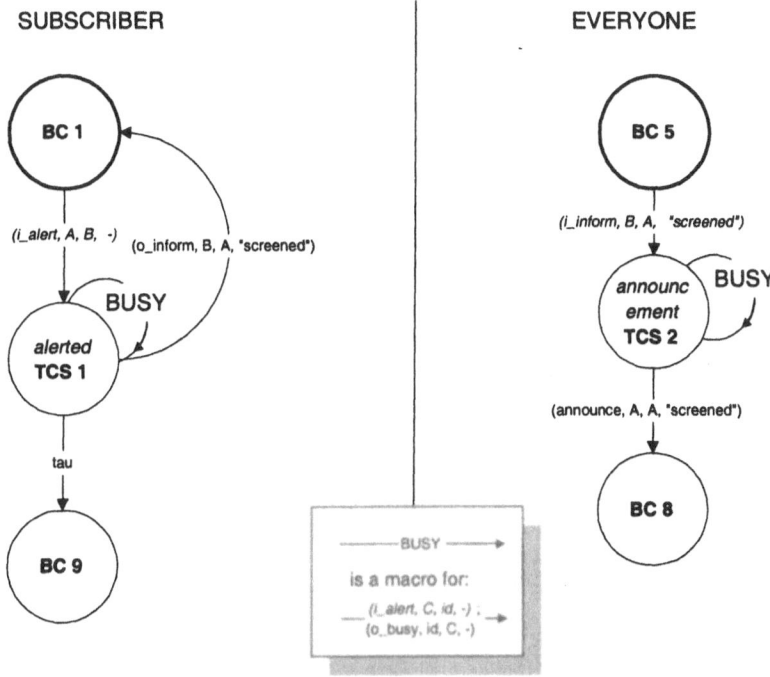

Fig. 2. Feature Terminating Call Screening [3]

The features, too, are given as labelled transition diagrams, with some BCM states and (usually) several new states (feature states). Feature integration is defined by adding or replacing, for a given basic call state, the transitions which are given in the feature definition. Here, a transition is replaced if the feature's transition has the same triggering event as the corresponding transition in the basic call; otherwise it is added. In the diagram for the feature Terminating Call Screening (Figure 2), for example, the transition from the idle state to BC9 of the basic call is replaced by the left diagram (for the subscriber), while in all basic call processes a new transition is added to the state BC5 ("o_busy?"). This transition is labelled with the message (i_inform,B,A,"screened") and leads to the new state TCS2.

The contest comprised the following features:

CFB Call Forward on Busy – divert calls when the line is busy

CNDB Calling Number Delivery Blocking – do not display the caller's number to the callee

CT Call Transfer – transfer an active call to another telephone

CW Call Waiting – allow subscriber to take a second call and switch between first and second call

GR Group Ringing – make three telephones ring for an incoming call to one of them, stop the ringing when one of them is answered

RBWF Ring Back When Free – ring back callers that got the busy tone (also known as Automatic Call Back)

RC Reverse Charging – callee pays for call

SB Split Billing – callee pays part of the call

TCS Terminating Call Screening – block calls from certain phones to the subscriber's phone

TL TeenLine – a PIN must be entered before calls can be made

TWC Three-Way Calling – allow the subscriber to initiate a second call and let all three parties talk with each other (also known as Conference Call)

VM Voice Mail – callers can leave a message if the phone is not answered (also known as CallMinder)

3   Observations

Analysing the feature definitions, we realized that the model given in the contest instructions was extremely prone to interactions. With a little practice we could anticipate many interactions by just looking at the features' diagrams. To explain how this worked, we classify interactions according to their causes:

1. one feature overrides trigger of the other feature (e.g. TCS & GR)
2. one feature bypasses trigger of the other feature (e.g. TCS & CW)
3. one feature sends a message that the other feature cannot process (e.g. TCS& TWC)
4. other causes (e.g. TCS & RBWF, RC & RBWF)

In the first two cases one of the features is not invoked when it should be; in the latter two, both features are active but interfere with each other in some way.

Interactions of type 1 are the easiest to detect: both features are triggered in the same (basic call) state by the same event. Due to the method of feature integration, the feature integrated later overwrites the transition introduced or altered by the earlier one. Interactions are usually serious because they are due to some fundamental incompatibility. In this case the interaction can only be resolved by limiting or refining the behaviour of one or both features.

Type 2 interactions are also not hard to recognize. Whenever a triggering message for one feature can occur in a feature state of another feature, this may lead to the former feature not being invoked even though it should. In the contest, all interactions of this type occurred because of the method of feature integration, and could clearly be avoided by a design that did not put so much emphasis on states. It is very likely that some of the type 2 interactions would then become type 4 interactions, i.e. the type 2 conflict indicated a serious interaction but for trivial reasons.

Interactions of the third type occur when a feature introduces a new message to the system, which may be received in a feature state of another feature. Since only basic call states are altered to be able to handle the new message, a feature on the receiving end will not be able to handle the new message. For these interactions, the same observations hold as for type 2: in a better, feature-oriented architecture, they would not occur.

The remaining interactions (type 4) are particularly interesting from the point of view of feature interaction detection. They indicate deeper problems in the way that features affect processing and distribution of information in the telephone network. There are no generic tests to detect such interactions, however they violate some 'sensible assumptions' about the working of the system. Furthermore, they only surface in the actual execution of the system, so they are not amenable to static analysis methods, such as proposed in the following section.

4   Static Analysis

The observations above prompted us to look for syntactic criteria to detect the first three classes of interactions. We introduce the following notation.

Let S be the set of all states (both basic call and all possible feature states), and E the set of all events (messages). To simplify the presentation we omit the subscriber parameters unless absolutely necessary.[1] Feature n is denoted by F_n; Sys may be a feature, or the Basic Call possibly augmented with a number of features.

$\text{Trans}(Sys) \subseteq S \times E$ one pair (s, e) for each transition in Sys, such that in state s, event e commences the transition

$\text{Triggers}(F_n) \subseteq S \times E$ the basic call states in which the F_n can be triggered, with the corresponding triggering event

$\text{Msgs}(Sys) \subseteq E$ all i_xxx and o_xxx messages that appear on transitions of Sys (ignoring the prefixes "o_" and "i_")

Note that $\text{Msgs}(Sys)$ contains information about all transitions, while $\text{Triggers}(F_n)$ only records the new or altered transitions from basic call states introduced by F_n. Let $\sigma(s, e) = s$ and $\varepsilon(s, e) = e$ denote the projections onto the first and second component, i.e. states and events, respectively.

---

[1] To be fully correct, one would need to take account of several variables, at least subscriber and partner, for each state.

Example 1. For the Terminating Call Screening feature we have:

$$\text{Trans(TCS)} = \{(\text{BC1}, \text{i\_alert}), (\text{BC5}, \text{i\_inform}),$$
$$(\text{TCS1}, \text{i\_alert}), (\text{TCS1}, \text{o\_inform}), (\text{TCS1}, \text{tau}),$$
$$(\text{TCS2}, \text{announce}), (\text{TCS2}, \text{i\_alert})\}$$

and

$$\text{Triggers(TCS)} = \{(\text{BC5}, \text{i\_inform}), (\text{BC1}, \text{i\_alert})\}$$

4.1 Interactions

With this notation, the following criteria characterise the first three classes of interactions given on page 166.

1. Later feature overrides earlier one:
   $\exists (s, e) \in \text{Triggers}(F_1) \cap \text{Triggers}(F_2)$
2. One feature bypasses a trigger of another feature:
   $\exists e \in \varepsilon(\text{Triggers}(F_1)) \cap \varepsilon(\text{Trans}(F_2) \setminus \text{Triggers}(F_2))$
3. F_1 may send a message which F_2 cannot handle ("Message not understood"):
   $\exists e \in \text{Msgs}(F_1) \setminus \text{Msgs}(F_2)$

These criteria will flag some potential interactions that *cannot* occur during normal execution of the featured system. In our experience with the contest model, however, they were quite accurate.

It is important to note that these criteria point to *causes* for interactions. There may be more than one actual, observable interaction for the same event or state-event pair satisfying one of the criteria.

Example 2. Since the Group Ringing feature, like TCS, is also triggered by an "i_alert" message in state BC1, we get a type 1 interaction:

$$\text{Triggers(TCS)} \cap \text{Triggers(GR)} = \{(\text{BCi}, \text{i\_alert})\}$$

Thus we can conclude, when TCS and GR are added to the same subscriber's telephone, whichever feature is added later will disable the earlier one, since it will override the triggers for the feature added earlier.

Furthermore we also get a type 3 interaction, since

$$\text{o\_inform} \in \text{Msgs(TCS)} \quad \text{but} \quad \text{i\_inform} \notin \text{Msgs(GR)}.$$

Note that a message "o_xxx" becomes "i_xxx" for the receiving phone, hence the third criterion says, *if* there is a situation in which the TCS feature sends an "o_notify" while the other phone is in an GR feature state, then that message cannot be processed (by the GR feature), leading to undefined behaviour. As our method stands at the moment, it is up to the user to check whether such a situation can actually arise. With these two features, the interaction can happen, in the following scenario. Assume that subscriber A has GR, and B and C are in the 'group'. If, for example B subscribes to TCS and has A on its screening list, then a call to A will result in an i_alert from A to B, which B will answer with (o_inform,B,A, "screened"). At this point A will be in a state introduced by Group Ringing, and will not be able to deal with that message.

Roughly 90% of the interactions we detected were discovered by applying these simple criteria. Yet, the interesting interactions are those that are not detected by these tests. The way that the contest model was set up, there were very few of these 'tricky' interactions, since most features clashed quite badly on the simple criteria. Assuming a good software engineering approach, though, which would take account of the 'easy' interactions and aim to avoid them in the first place, the 'tricky' ones, i.e. those not detected by our syntactic criteria, become crucial. At this point, simulation, testing and model checking come into their own again, because the interactions that cannot easily be detected by syntactic criteria are likely to show up only in longer runs of the system.

5   Modelling the system in CSP

5.1   A Network of Basic Call Processes

Modelling a single basic call in CSP is very simple: the states become CSP processes and all messages become events, the send and receive operations fit very nicely. A transition with multiple messages is split up into a sequence of events.

The problems start, however, when one composes several such basic call processes into a network. Now, the different processes may 'de-synchronize' since there is nothing to enforce the atomicity of the transitions in the diagram. Hence one line could embark on a transition if this was triggered by a local event and thereby prevent another line from sending a message to it. In other words, the naive, literal translation from the contest instructions allowed some internal choice between transitions that need to synchronize with other transitions, leading to possible deadlocks.

It turned out, though, that this problem was not too hard to solve: it was necessary to reorder the events labelling each transition, so that all external communication events came first on the respective transitions. Eventually, all transitions started with one input or output event, if they contained one at all, followed by only internal (local) events.[2] In CSP terms this meant offering all events that the line processes had to synchronize on in external choice constructs.

---

[2] For some features this meant introducing extra states and transitions, since some of them used multiple send operations in one transition.

```
----------------------------------------------------------------
-- Feature Terminating Call Screening
----------------------------------------------------------------
-- This is designed for a maximum of four phones: any phone above C
-- gets $TCSSetD as its screening set (only relevant if in subscriber
-- set $TCSSubs).
----------------------------------------------------------------
$TCSSubs = {C}
$TCSSetA = {}
$TCSSetB = {}
$TCSSetC = {A}
$TCSSetD = {}

screen(A) = $TCSSetA
screen(B) = $TCSSetB
screen(C) = $TCSSetC
screen(_) = $TCSSetD

BCM(state:{idle},x,x) += if member(x,$TCSSubs)
                           then i.m_alert?y:Lx({x})!x!none -> TCS1(x,y)
                           else BCM(idle,x,x)

TCS1(x,y) =    (if member(y,screen(x))
                  then o.m_inform.x.y.screened -> BCM(idle,x,x)
                  else BCM(t_busys,x,y))
               [] i.m_alert?z:Lx(x)!x!none -> o.m_busy.x.z.none -> TCS1(x,y)

BCM(state:{o_busyp},x,y) += i.m_inform.y.x.screened -> TCS2(x,y)

TCS2(x,y) =    announce.x.tcs -> BCM(o_wait_onhook,x,y)
               [] i.m_alert?z:Lx(x)!x!none -> o.m_busy.x.z.none -> TCS2(x,y)

-- Observer process: after x has rung y ("alert.y.x"), CONNECT(x,y)
-- blocks all messages "alert.s.t" (t is ringing s), thereby eventually
-- causing a deadlock of the whole system.
CONNECT(x,y) = (alert.y.x -> STOP) [] (alert?s?t -> CONNECT(x,y))

-- deadlock occurs: A is allowed to connect to B ==> CONNECT deadlocks
assert CONNECT(A,B)[|{|alert|}|]POTS :[deadlock free [F]]
-- no deadlock: TCS prevents the message "alert.C.A" occurring
assert CONNECT(A,C)[|{|alert|}|]POTS :[deadlock free [F]]

-- end of feature -----------------------------------------------
```

BCM(state,subscriber,partner) denotes a basic call state,
$TCSSubs is the set of subscribers to TCS (a parameter of the feature),
$TCSSetA through **$TCSSetD** represent the screening lists of the respective subscribers (feature parameters).

Fig. 3. CSP_M^{FC} code for Terminating Call Screening

5.2   A Feature Construct for CSP

To automate feature integration we extended the CSP_M syntax with a simple feature construct for the textual representation of the feature diagrams. (CSP_M is the subset of CSP accepted by the FDR tool.) We will denote the extended language by CSP_M^{FC}.

A feature definition may contain any number of standard CSP_M definitions. These will simply be added to the base system, and the user is responsible for avoiding name clashes (re-definitions).

On top of plain CSP_M, there are two new constructs:

- Transitions can be added or replaced by means of special definitions with the following syntax:
  $process(p_1, \ldots, p_n)$ += *new definition*
  where *process* is a process name from the base file. The actual parameters can be 'captured' and are referred to by $p_1, \ldots, p_n$ in the *new definition*. It is also possible to restrict the range of parameters in certain cases.[3]
- Feature variables or parameters can be defined. Any name prefixed with $ is interpreted as a feature variable. Such variables can be assigned values in the feature file or on the command line when invoking the integrator (see below). All (non-defining) references in the feature file are textually replaced by the value thus given.

5.3   Feature Integration

We wrote a Python script (which we call the "feature integrator") which combines a feature definition (written in CSP_M^{FC}) and a base system in a CSP_M file and produce a new CSP_M file defining the featured system. Features may be parametrized, and the feature integrator allows us to set values for the parameters at integrate time, overriding any default values given in the feature definition.

The current prototype relies quite heavily on syntactic conventions in our POTS model, e.g. += definitions only work for BCM states and the number of parameters is fixed. On the other hand this enabled us to substitute sensible defaults for the 'don't care' place holder ($\_$).

After integrating a feature, the resulting CSP_M file can be used as the base file for further feature integrations or, of course, be analysed using PROBE and FDR2 [4] [5].

5.4   Detecting Interactions

To detect feature interactions in the CSP_M model, we applied several techniques.

First we used FDR2 to check the featured systems for deadlock. Obviously the telephone system should never deadlock – it must always be possible for every line to get back to the initial state. However, since we were working on a system with four phones, we could not detect local deadlocks, i.e. situations

---

[3] This is rather an *ad hoc* solution to deal with the specific form our POTS model takes.

[4] www.formal.demon.co.uk/FDR2.html

in which one or more phones had no more enabled transitions but where there was at least one which could still move – even if that only meant going offhook and onhook repeatedly. If FDR2 allowed the user to impose fairness constraints, such situations could be detected.

Secondly, we explored the behaviour of the system using PROBE, to test if the featured system *could* behave in the intended way. PROBE allows the user at every step to choose one of the enabled events and thus to simulate a run of the system. Hence it cannot be used to verify the absence of undesirable behaviour.

This is where FDR2 comes in again. While the previous two techniques are mainly for debugging, FDR2 can explore all possible executions of a system. The central method used for this is refinement checking in one of three models.[5] However, since features both add and remove behaviours, refinement is not such a useful notion. In general, the featured system will not refine the base system, nor vice versa.

Instead we coded desirable or undesirable patterns of behaviour as observer processes and composed them with the system in question, synchronising on the events that were relevant for the behaviour we wanted to test for. The observer processes were designed in such a way that the presence of the behaviours they represented lead to a deadlock in the composite system. This allowed for exhaustive checking of properties.

An example observer process can be seen in figure 3: the process CONNECT(x,y) monitors the "alert" messages of the telephone system. CONNECT(A,B) initially accepts all "alert" messages but after an alert.B.A event it refuses all further "alert" messages. The message alert.B.A indicates that phone B is ringing and that A is the caller. Hence once A has rung B, no other phone in the whole system can ring, and so a deadlock ensues. Conversely, no deadlock exists in the system if B subscribes to TCS and A is on B's screening list. (In the example of figure 3, the first assertion is false, for B does not subscribe to TCS, the second assertion is true, since C screens calls from A.)

Please note that it would not suffice for CONNECT(x,y) to refuse alert.y.x: in this case, the remaining phones could still perform transitions indefinitely, and no deadlock would be detected.

5.5   Limitations

The expressive power of observer processes in CSP is rather weak. Unlike 'never claims' in SPIN/PROMELA, which uses Büchi automata to deal with infinite behaviours (e.g. liveness properties, in general recognition of ω-regular traces), observer processes in CSP can only be used to detect the presence of finite traces.

---

[5] Traces, failures and failures-divergence model.

As with SPIN, some observer processes led to a huge blow-up in the state space, which made it impossible to verify the properties.[6]

Another drawback of using observer processes is that they need to be coded by hand which is a rather error-prone procedure. Contrast this with expressing a property in temporal logic, with subsequent automatic translation to the corresponding Büchi automaton.

This list of drawbacks might give the impression that the model-checking approach is deeply flawed. However, we would like to point out that the fact that a simple static analysis detected such a large proportion of the feature interactions hinged on the specific model the contest defined. Also, the expressiveness of model-checking languages varies greatly, with regards to both the description of models and the properties that can be checked. Indeed we would argue that model checking is still an invaluable tool in proving the absence of unwanted behaviour, and in finding deep errors.

6   Feature Interactions

A full list of the (two-way) interactions we detected among the twelve features of the contest, with explanations can be found in [4]. Table 1 may give an impression of the sheer number of interactions that we found. We only elaborate on a small selection here.

	CFB	CNDB	CW	RBWF	RC	SB	TCS	TL	TWC	VM	CT	GR
CFB	××	–	–	–	–	–	–	–	–	–	–	–
CNDB	××		–	–	–	–	–	–	–	–	–	–
CW	××	×	×	–	–	–	–	–	–	–	–	–
RBWF	×××××	××	×××	×	–	–	–	–	–	–	–	–
RC	××			××	–	–	–	–	–	–	–	–
SB	××			××	×	–	–	–	–	–	–	–
TCS	×	×	×	×××			–	–	–	–	–	–
TL	×			××			×	–	–	–	–	–
TWC	×××	×	×××	××××	×	×	×	×		–	–	–
VM	×		××	××	×	×	×		××	×	–	–
CT	××××	×	×××	××××	×	×	××	××	×××	××	××	–
GR	××	×	×××	×××	×	×	××		××	××	××	×

Each × in the table stands for a distinct interaction.
We did not distinguish feature combinations by the order of feature integration, hence the top right half of the table is not used.

Table 1. Feature Interactions Phase 1

---

[6]   ... at least with the computing resources available to us.

- **TCS & GR:** Both features are triggered by an incoming alert message in the idle state (type 1 interaction). Therefore they cannot both be active on the same line.
- **TCS & CW:** When the a subscriber of Call Waiting is in a call, further incoming calls (i_alert message) are not screened, since TCS is triggered only in the idle state; this is a type 2 interaction.
- **TCS & TWC:** A typical type 3 interaction occurs because Terminating Call Screening introduces a new message (i_inform). All lines in the network are upgraded so that they can deal with this new message, but only in BC states. So if someone uses Three Way Calling to call a Call Screening subscriber, an 'i_inform' message from that line to the (TWC) caller cannot be processed by the TWC feature.
- **RBWF & RC:** Ring Back When Free initiates the ring-back without a dial message, therefore Reverse Charging will not be triggered. This is a type 4 interaction, however, if the Ring Back feature were redefined to use the dial message, we would still get an interaction, namely of type 2.

7   Conclusions

Taking part in the contest was fun, but also quite exasperating at times, due to ambiguities in the contest instructions and bugs in the specifications of POTS and the features. We spent a lot of time on getting our model to work at all – many problems were synchronisation problems and mismatches of sending and receiving, i.e. bugs in the protocol.

The verification task was made difficult by the lack of a clear description of what was considered incorrect or undesirable behaviour on the one hand, by certain shortcomings of the model checker FDR2 on the other. The former is probably quite realistic, because at specification time, the requirements are often not fixed, and only in the process of "playing with the system" does one discover contradictory requirements, or additional assumptions that need to hold.

We faced the usual problems of underspecification and ambiguity when dealing with a natural language description. For example, what constituted a correct billing record was not stated. The CNDB feature relied on some internal bookkeeping in the switch, which was never made explicit. However, to assess the effects of 'anonymous' messages, one needs to make assumptions about the capabilities of the switch in this respect. It is unlikely that all contestants will make the same assumptions, hence it becomes almost impossible to compare their results about this feature.

While the success of the syntactic method described in section 4 seems to cast doubt on the value of model checking for the detection of feature interactions, we would not have reached an 'implementation' of POTS and the features without the aid of model checking. If PROBE and FDR2 had been nothing more than debugging aids in the development of our system, they would have been invaluable in getting the system right, and moreover gaining a good understanding of it.

Furthermore, we believe that the syntactic method captures mainly the 'obvious' interactions. Once these are out of the way, one needs a way to find those interactions that can result, e.g. from different interpretations of the same data in different features. These interactions may result in strange behaviour or simply in the violation of invariants, but this will only become visible in runs of the system (model).

Acknowledgments. Thanks to anonymous referees for helping us to improve the paper. Financial support from the EU through Esprit working groups FIREworks (23531), and from British Telecom is gratefully acknowledged.

References

1. M. Calder and E. Magill, editors. *Feature Interactions in Telecommunications and Software Systems VI.* IOS Press, 2000.
2. M. Kolberg, E. Magill, D. Marples, and S. Reiff. Results of the second feature interaction contest. In Calder and Magill [1], pages 311–325.
3. M. Kolberg, E. Magill, D. Marples, and S. Reiff. Second feature interaction contest. In Calder and Magill [1], pages 293–310.
4. M. C. Plath and M. D. Ryan. Entry for FIW'00 Feature Interaction Contest. Technical report, School of Computer Science, University of Birmingham, February 2000. Available from `ftp://ftp.cs.bham.ac.uk/pub/authors/M.D.Ryan/00-fiw-contest.ps.gz`.
5. A. W. Roscoe. *The Theory and Practice of Concurrency.* Prentice Hall, 1999.

Stack Service Model

D. Samborski

LORIA, BP 239, 54506 Vandœuvre-lès-Nancy, France.

Abstract. Stack Service Model (SSM) is an architecture that describes the telecommunications world. Services are plug-and-play boxes associated with their subscribers' phones. The phones contain a stack structure that ranges services by priority. The communications between stacks are handled by token (message) exchange. We propose an appropriate specification language. The entire system is verified by model-checking in order to detect feature interactions. A Feature Interaction Manager can be built with SSM – we describe a method of resolution of interactions and an abstract specification method that improves communication with the customer.

1 Introduction

In their survey on the feature interactions problem in telecommunications systems, Keck and Kuehn [14] remark that "In the detection category, there is an overwhelming number of formal method approaches, using all kinds of methods, while there is very little support for real implementations (. . .)". We propose an architecture that can be adopted for realsize systems – telephones and services communicate by token (message) exchange. The services are placed in stacks associated with phones. Such a choice (already existing, see [2, 3, 13, 18]) presents several advantages:

- it makes service composition simple and modular – services are plug-and-play boxes;
- it avoids some interactions by assigning priorities to services in the same stack;
- it allows a fine analysis of events – everything is controlled by token exchange and every token can be analysed;
- token exchange can be done in any appropriate way, a structure similar to the Internet can be used.

Moreover, the distributed nature of the network is a step towards the future Internet-like telephony (see [15]). It also helps to adopt our techniques to systems other than telephonic ones.

Without creating a new architectural concept, we introduce a simple and useful specification language that describes token processing. A tool support is provided for service validation, animation of scenarios and interaction detection. We used SSM and its tools to participate in the Second Feature Interactions Contest [16].

This work is the result of a collaboration with prof. G.-C. Roman (Washington University in St.Louis, MO, USA). The author wishes to expresses his gratitude to Professor Roman for his precious help.

The paper is organized as follows: first we present the SSM architecture and the structure of services. The SSM specification language is introduced. In the section 3, we show how to compose features and how to detect feature interactions. The section 4 focuses on futher development of SSM – building token supervisors that resolve interactions and an abastract specification method familiar to engineers, that helps to communicate with the customer in a more efficient way.

2   Description of SSM

2.1   Telecommunications World

The telecommunications world contains three major types of actors. **Users** initiate activity in the network; **devices** that interface connecting users, and a **network** that allows the physical linking among devices. Most of the devices are phones, others can be more special like, for example, a billing machine. Every phone has the POTS service that allows it to call and to receive calls. Other services are intended to modify the POTS behaviour – they can even cancel entirely the POTS' functionality. Note that we don't distinguish features from services.

In SSM, services are placed inside the phones. This is the most important difference between SSM and DFC [13] where phones and services are separated. In SSM one should always speak about *instances of services*. When a user subscribes to a service, he receives a "black box" that is added to his phone. A phone contains a stack structure that holds the service boxes. The stack structure defines the order (priority) among services of the same phone. Users can change priorities by reordering services but this may result in a wrong service behaviour. For many services however the order has no importance; one cannot establish a priority relation among them. Note that potentially POTS can be placed anywhere in a stack. In our actual implementation its place is always the lowest one, but one can imagine services that require tokens already processed by POTS from its own stack.

Figure 1 shows the interconnection of two phones. The grey arrows correspond to token exchanges. Tokens are formatted messages that contain some mandatory information (the type of token, the originating and the destination phone addresses) and any other optional information (voice messages to be played back, other telephone numbers, time records, etc). Token exchange between phones is always done via the network. The network distributes tokens according to their destinations. We don't assume anything about how the tokens circulate in the network – they can be delivered in any efficient way, for example using sockets in a computer network.

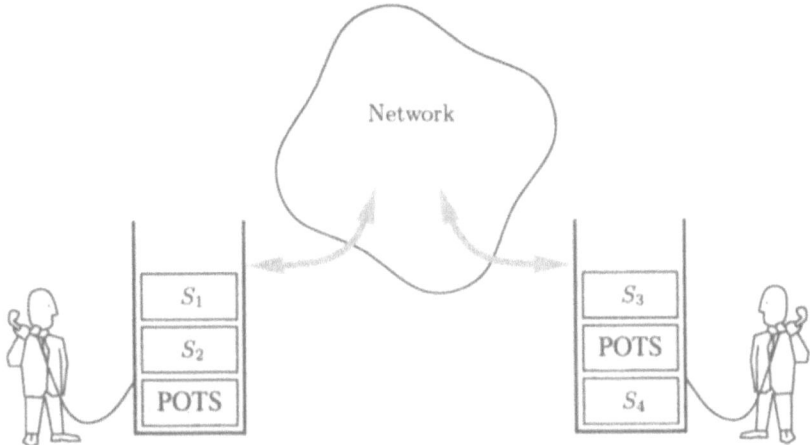

Fig. 1. Two communicating phones

All activities are controlled by token exchanges. The user acts on phones by placing user actions whose result is a creation of *user-tokens* addressed to his phone. The phone reacts by sending *system-tokens* to appropriate stacks. There is a special stack, called *System stack* or SYS, whose role is double: a) it handles system events, like sending tones to telephones, establishing connections, etc; b) it contains services that are not associated with any particular phone, e.g. credit card calling. Some services can have two separate parts – one in the subscriber's stack and one in the SYS. The system stack has the same structure as any other stack. It can be located anywhere outside the network.

2.2 Token Processing

A token arriving at a stack goes down passing successively through the services inside. A service can retrieve a token according to its type; it can absorb it and/or send new tokens to the services below or outside the stack. A service can also modify stack variables while processing tokens. Stack variables are used to describe the state of the stack. Newly added services add their variables to the set of stack variables.

Since the higher placed services receive down coming tokens before the others, they have a higher priority. Thus, the position in the stack defines a priority relation among services.

To avoid the same variables to be modified at the same time by different token being processed, we allow only one token to go down the stack at a time. Other tokens arriving to the stack are placed in a queue, called **synchronizer**, to wait for their turn. In fact, the synchronizer has two distinct queues – a *S-queue* for system-tokens and a *U-queue* for user-tokens. These two queues are handled differently. After placing an action, the user waits for the system reaction. If he places a second action before the system can complete the processing of the first one, this action is taken in account only after the processing of the first action is done. That is why, we supply the U-queue with a door (a *U-door*) that opens when the telephone is ready to receive user tokens and is closed the rest of the time. The use of a door permits also to cope with token delivery delays in real systems.

There are three synchronizer rules that define when a token can be let into the stack. In all cases, the stack shouldn't contain tokens moving down, in other words, the stack should be empty of tokens.

1. If the U-queue is not empty and the U-door is open, then let the first U-token in and close the U-door;
2. If the U-queue is empty and the U-door is open, then let the first S-token (if any) in the stack. The U-door stays open if it isn't closed explicitly;
3. If the U-door is closed, then let the first S-token (if any) in the stack.

When the U-door is open, user-tokens have priority over system-tokens. The decision about opening the U-door is taken by services (there is an explicit command to open/close the U-door).

2.3   Semantics of SSM

Let us describe now the operational semantics of SSM. A stack is characterized by a structure of the form:

$$\{n, \ L_u, \ L_s, \ \varepsilon, \ d, \ S_1 \cdot S_2 \cdots S_k, \ V\}$$

- n (name) is the ID of the stack (we use capital letters);
- L_u (user line) is the U-queue whose structure is FIFO. Adding a token to the queue is noted $L_u \downarrow t$. $hd(L_u)$ and $tl(L_u)$ are respectively the head and the tail of the queue;
- L_s (system line) is the S-queue ;
- ε is the state of the U-door: open (1) or closed (0) ;
- d is the down coming token. If no token comes down, we note δ;
- $S = S_1 \cdot S_2 \cdots S_k$ is the set of services in the stack. The order is important – the first one is the highest placed;
- V is the set of pairs variable/value for the stack.

Tokens in the network form the set P.

We define semantic rules using the operator $\longrightarrow$ which acts inside the space of states. It can be interpreted as "at the next moment". The functions *type*, *orig*, *dest* and *mess* give the corresponding token component.

Token release. Users can release U-tokens to their phones. There are four U-tokens for the moment: **offhook**, **dial**, **flash** and **onhook**.

$$\frac{type(t) \in T_{user}}{(\{n,\ L_u,\ L_s,\ \varepsilon,\ d,\ S,\ V\},\ P) \ \longrightarrow \ (\{n,\ L_u \downarrow t,\ L_s,\ \varepsilon,\ d,\ S,\ V\},\ P)}$$

Get tokens. A stack retrieves tokens that are sent to it:

$$\frac{(t \in P) \wedge (n \ = \ dest(t))}{(\{n,\ L_u,\ L_s,\ \varepsilon,\ d,\ S,\ V\},\ P) \ \longrightarrow \ (\{n,\ L_u,\ L_s \downarrow t,\ \varepsilon,\ d,\ S,\ V\},\ P \setminus \{t\})}$$

The tokens are first stocked in the synchronizer.

Let a token in. Using the above described algorithm to manage the U-door, we obtain three rules:
The first case: the U-door is open and there are U-tokens in the U-queue.

$$\frac{t \ = \ hd(L_u)}{(\{n,\ L_u,\ L_s,\ 1,\ \delta,\ R,\ V\},\ P) \ \longrightarrow \ (\{n,\ tl(L_u),\ L_s,\ 0,\ t,\ S,\ V\},\ P)}$$

The second case: the U-door is open but there is no U-tokens in the queue. The S-tokens can be let in.

$$\frac{t \ = \ hd(L_s)}{(\{n,\ \emptyset,\ L_s,\ 1,\ \delta,\ R,\ V\},\ P) \ \longrightarrow \ (\{n,\ \emptyset,\ tl(L_s),\ 1,\ t,\ S,\ V\},\ P)}$$

The third case: the U-door is closed. We let the S-tokens in.

$$\frac{t \ = \ hd(L_s)}{(\{n,\ L_u,\ L_s,\ 0,\ \delta,\ R,\ V\},\ P) \ \longrightarrow \ (\{n,\ L_u,\ tl(L_s),\ 0,\ t,\ S,\ V\},\ P)}$$

Process tokens. A token coming down the stack is processed by services. As the result of the processing, new tokens can be released and variables can be re-affected. Hence, the result of the processing is a triplet $\{d, U, W\}$ where d is the new down coming token, U is the set of newly released tokens sent outside the stack and W is the new set of variable/value pairs. We must also add the U-door state γ as it can also be changed by token processing. We write for a service R:

$$R(t,\ \varepsilon,\ V) \ = \ \{d,\ U,\ \gamma,\ W\}$$

If the service R doesn't process the token d, then $R(t,\ \varepsilon,\ V) \ = \ \{t,\ \emptyset,\ \varepsilon,\ V\}$.

$$\frac{R(t,\ \varepsilon,\ V) \ = \ \{d,\ U,\ \gamma,\ W\}}{(\{n,\ L_u,\ L_s,\ \varepsilon,\ t,\ (R \cdot Q),\ V\},\ P) \ \longrightarrow \ (\{n,\ L_u,\ L_s,\ \gamma,\ d,\ Q,\ W\},\ P \cup U)}$$

The lowest situated service send all the tokens out from the stack.

$$\frac{(R(t,\ \varepsilon,\ V)\ =\ \{d,\ U,\ \gamma,\ W\}) \wedge (R\ =\ tl(S))}{(\{n,\ L_u,\ L_s,\ \varepsilon,\ t,\ R,\ V\},\ P)\ \longrightarrow\ (\{n,\ L_u,\ L_s,\ \gamma,\ \delta,\ \emptyset,\ W\},\ P \cup U \cup \{d\})}$$

A typical token processing is a succession of the above described rules.
Below we introduce some useful notations. When a token t arrives to the
synchronizer queue of a stack A, we say that the token *arrives at the stack*
and we denote the fact by $t \triangleright A$. When a token t moving down the stack
arrives at a service S (that means that the upper adjacent service has sent
it down or S is the highest service and the token comes directly from the
synchronizer), we say that t arrives to S.

When a service sends a token t outside the stack A, the token leaves the
stack to the network. We denote this fact by $A \triangleright t$. We use this notation in
specifying safety properties. Here are two examples:

- The Originating Call Screening forbids A from calling B. When a telephone calls another one, it sends a token of type **alert**. The safety property is then:

$$\neg(A \triangleright \textbf{alert}(A, B)) \tag{1}$$

- The Terminating Call Screening prevents B from being called by A. The
  telephone starts to ring when the token **alert** enters POTS. The safety
  property is:

$$\neg(\textbf{alert}(A, B) \triangleright POTS(B)) \tag{2}$$

The notions *leaves a stack* and *arrives at a stack* are closely linked (if the
network can assure token delivery): a token $\mathbf{t}(A, B)$ leaves the stack A and
arrives at the stack B. One cannot say anything about arriving at services
since higher located services can block or change the down coming token.

2.4   Token types

After having specified 15 services we can make a list of the tokens we used.
These tokens are:

U-tokens:
offhook(A, A), **dial**(A, A, B), **onhook**(A, A) and **flash**(A, A) do what they
are supposed to do. We authorize them to be released at any moment. Service
specification will choose only those that change the actual state of the system.

S-tokens:
alert(A, B) demands for a call from A to B.
abandon(SYS, A, B) notifies A that B has abandoned. Starts the end tone.

billing_offhook, billing_onhook, billing_start, billing_stop manage the payment.

busy_line(SYS, A, B) notifies A that B is busy. A is the one who has asked for a call.

busytone(A, SYS, B) notifies B that A is busy. B is the one who has asked for a call.

change_callee(A, B, C) notifies that the called party for B is no longer A but C.

connect(A, SYS, B) accepts the connection required by B.

connection(SYS, A, B) notifies A that the required connection to B is accepted.

dial_tone(SYS, A) starts the dial tone on A.

disconnect(A, SYS, B) notifies B that A has abandoned.

ringtone(A, SYS, B) notifies the calling party B that the phone A starts to ring.

free(SYS, A, B) notifies the calling party A that the called party B is ringing.

requireDialTone(A, SYS) asks for the dial tone.

stop_alert(A, SYS, B) notifies B that A has abandoned the calling attempt.

stop_ringing(SYS, A, B) asks A to stop ringing for the call required by B.

We see that many tokens are sent from A to B via the system stack SYS. This allows system-based services to control basic token exchanges. Here is the list of standard token transformations by SYS:

$$
\begin{aligned}
\textbf{requireDialTone} &\longrightarrow \textbf{dial\_tone} \\
\textbf{busytone} &\longrightarrow \textbf{busy\_line} \\
\textbf{ringtone} &\longrightarrow \textbf{free} \\
\textbf{connect} &\longrightarrow \textbf{connection} \\
\textbf{disconnect} &\longrightarrow \textbf{abandon} \\
\textbf{stop\_alert} &\longrightarrow \textbf{stop\_ringing}
\end{aligned}
$$

Services can introduce their own token types although this is not recommended. Service designers should respect the standard set of token types if they want that their services understant the others. Sometimes, however, introducing new token types is necessary. In our case study, only the services Voice Mail and Group Ringing introduce their own tokens (that are not supposed to be used by other services).

2.5 SSM Specification Language

SSM is also the name for a simple specification language. This language is basically designed to handle token exchanges – retrieve tokens, create new tokens and send them back and forth. It has the common features of any programming language – variable assignments, control structures, etc.

The main operators of SSM are GET and SEND. GET retrieves the token of certain type passing through the service, i.e., it defines a filter for tokens. SSM specification of a service is a set of such a filters. The general structure of a service is

> SERVICE S
> VARIABLES
> $\quad v_1 \; := \; a, \;\; v_2, \; \ldots$
> INVARIANT
> $\quad pr_1 \wedge pr_2 \wedge \; \ldots$
> GET **t1** $\{ \; expr_1 \; \}$
> $\quad \vdots$

Service variables are local to the stack. They can be initialized or not. The invariant is a conjunction of state-based predicates expressed in predicate logic. They can also use the "arrives" notations introduced before. The invariant preservation is verified every time a token enters the stack, a service in the stack or when a token leaves the stack. GET filters enclose expressions that define token processing for different token types. Usually, a service processes only few types of tokens. A token that doesn't match any type processed by the service passes through the service unchanged to the next service in the stack or it is sent outside the stack if there is no other service below.

Here is an example of Call Forwarding on Busy Line service.

```
SERVICE CFBL
GET alert(A, THIS){
    IF busy = true THEN
        SEND alert(A, forward_number) UP
        SEND billing_forward(THIS, SYS, A, forward_number) UP
    ELSE
        SEND alert(A, THIS) DOWN
    END
}
```

The service reacts on receiving the **alert** token (that means that the phone receives a call). The token comes from any phone A and is destinated to the current phone, denoted by the keyword THIS. The filter verifies whether the subscriber is busy by checking the correspondent variable. If it is the case, it sends a new **alert** token making a forwarded call. Otherwise, it lets the token pass through, eventually to POTS. The name of the variable *forward_number* is explicit. The service specification is not complete since we didn't say anything about the invariant and the variables of CFBL. This service introduces only one variable – *forward_number* (and uses the variable *l usy* of POTS). As for the invariant, the next subsection describes how to construct it.

2.6   Find out Safety Properties

Interactions detection is based on verifying the preservation of services' safety properties (for the moment we don't deal with liveness properties), that is we are looking for "bad" interactions. This places our method in the category of General Properties approach, as classified by Bredereke [5]. Thus, we face the common difficulties of this approach – how one can be sure that the set of given properties is complete and reflects what we really want the service to do? SSM concepts give some hints about finding out these properties.

Anything that can happen to a phone is the result of token processing. The invariant of a service S can be violated if a) the service S has processed a token driving the phone into a "bad" state; b) another service R in the stack has processed a token driving the phone into a state that violates the invariant of S. When specifying different services (those of the Second Feature Interaction Contest [16]) we realized that the number of different token types is quite limited – about two dozens. What we propose is a heuristic method, simple yet very useful.

We propose to verify explicitly for every token type, whether a token can somehow be harmful for a service.

Let us come back to the example of Call Forwarding service. The purpose of the service is to forward calls when the phone is busy and let them in when the phone is idle. The first requirement can be translated into a safety property:

$$(busy = \text{true}) \Rightarrow \neg(\textbf{alert}(A, \text{THIS}) \triangleright POTS(\text{THIS}))$$

for any phone A. Similarly, the second requirement corresponds to a property:

$$(busy = \text{false}) \Rightarrow \neg(\text{THIS} \triangleright \textbf{billing\_forward}(\text{THIS}, SYS, A, forward\_number))$$

Now we should analyse which tokens can be harmful for the service. We found only one of them: the token **busytone** should never be sent

$$\neg(\text{THIS} \triangleright \textbf{busytone}(\text{THIS}, SYS, A))$$

for any $A \in Phones$. By making a disjunction of the last three predicates, we obtain the invariant of CFBL. The service specification is now complete.

Another example is the Originating Call Screening service whose role is to prevent its subscriber A from calling the phone B. The calling action is translated into sending an **alert**(A, B) token – by checking that A doesn't send it, we obtain a safety property (see our example (1)). This is not, however, the only safety property we can supply. The following depends on what we want from the OCS service – either we want that A doesn't *dial the number of B* (because the call is too expensive), or we want A not to *be connected* *to B* (because *B* is an adult-only line). In the first case it is sufficient to add the requirement that the user token **dial**(THIS, THIS, *B*) doesn't reach POTS service of the stack *A*. In the second case one must analyze the token exchange that occurs once *A* connects *B* – thus we obtain a set of new safety properties:

$$\neg(\text{THIS} \triangleright \textbf{alert}(\text{THIS}, B))$$
$$\wedge \neg(\text{THIS} \triangleright \textbf{billing\_start}(\text{THIS}, SYS, B))$$
$$\wedge \neg(\textbf{connection}(SYS, \text{THIS}, B) \triangleright OCS(\text{THIS}))$$
$$\wedge \neg(\textbf{set\_tone}(B, \text{THIS}, \text{"talking"}) \triangleright POTS(\text{THIS}))$$

Suppose that *C* is a Call Transfer subscriber. He can make *A* join *B* without sending an **alert** token between them. If *A* has the OCS service forbidding him to call *B*, the protection may be by-passed easily – *A* can call *C* and ask him to join *B*. This cheating will, however, be immediately detected because connection tokens need to be passed between *A* and *B*. If we were using only the property (1), this interaction would not be detected.

3   Feature Composition

To compose services in SSM one simply inserts service boxes in the right places on stacks. In fact, we compose *instances* of services and not the services themselves. Different interactions can occur depending on the mutual disposition of service boxes. If we want to add a service we first test all possible combinations of this service with other services in the system. With every new instance of a service, new combinations become possible, so we need to re-test the whole system. This sounds terrible, but the reality is not so cruel. For each service there exists a maximum number of instances beyond which, no new interactions can appear.

Let us denote by $S_1 \oplus S_2$ the composition of instances of services S_1 and S_2. Since the composition includes all possible combinations between the instances, the operator $\oplus$ is commutative and associative.

3.1   Interacting Groups

Non Interacting Cases Let $T_{in}(S)$ be the set of tokens that can be processed by the service *S*. It is defined by the set of types $\Theta(S) = \{\theta_1, \theta_2, \dots, \theta_n\}$ used in the GET filters of *S*:

$$T_{in}(S) = \{t \in \textit{Tokens} | \exists \theta \in \Theta(S).type(t) = \theta\}$$

By processing tokens from $T_{in}(S)$, the service *S* sends back and forth new tokens that form the set $T_{out}(S)$. Formally, the result of processing of a token *t* by the service *S* can be expressed by:

$$S(t, \varepsilon, V) = \{d, U, \gamma, W\}$$

where ε and γ are the states of the U-door before and after the token processing, V and W are the sets of pairs variable/value of the stack, d is the resulting down coming token (if any) and U is the set of tokens leaving the stack immediately after processing. The set $T_{out}(S)$ is defined as follows:

$$T_{out}(S) \;=\; \bigcup_{t \in T_{in}(S)} \{d\} \cup U$$

Consider the following system of conditions

$$\begin{cases} T_{in}(S_1) \cap T_{in}(S_2) \;= \emptyset \\ T_{in}(S_1) \cap T_{out}(S_2) = \emptyset \\ T_{in}(S_2) \cap T_{out}(S_1) = \emptyset \end{cases} \tag{3}$$

The first condition means that the services S_1 and S_2 never compete for the same tokens. The two last ones say that any service doesn't process the tokens issued from another service. We can now assert that S_1 and S_2 are not related between them by any priority. Their mutual position can not be a source of interactions.

To be sure that S_1 and S_2 don't interact at all, we must check that every service doesn't send a token that can violate the other's service invariant. Let $J_{in}(S)$ be the set of tokens such that for any $j \in J_{in}(S)$ there exists a safety property in the invariant of S of the form:

$$\neg(j \triangleright S) \quad \text{or} \quad \neg(j \triangleright \text{THIS})$$

We define the set of tokens that should not leave the stack $J_{out}(S)$ in the same way. Now we can require that:

$$\begin{cases} J_{in}(S_1) \;\cap T_{out}(S_2) = \emptyset \\ J_{in}(S_2) \;\cap T_{out}(S_1) = \emptyset \\ J_{out}(S_1) \cap T_{out}(S_2) = \emptyset \\ J_{out}(S_2) \cap T_{out}(S_1) = \emptyset \end{cases} \tag{4}$$

The first two equations means that a service doesn't send a token that violates directly the invariant of another service located in another stack. The last two ones mean the same thing but in the case where both services are located in the same stack.

The last thing to control is the variable assignment. If in addition to (4) a token sent by S_1 doesn't provoke a modification of variables that appear in the invariant of S_2, then it cannot violate the invariant. This condition is true if the system 3 is satisfied since tokens sent by S_1 are not processed by S_2 and so cannot modify its variables. Meanwhile, we should assure that *all* the tokens accepted by S_1 don't violate the invariant of S_2 and vice-versa – in the case where S_1 and S_2 are situated in the same stack. Let $V(S)$ be the set of variables that appear in the invariant of S and $W(t, S)$ – the set of variables modified after the service S has processed the token t. We require that:

$$\begin{cases} \bigcup_{t \in T_{out}(S_1)} W(t, S_1) \cap V(S_2) = \emptyset \\ \bigcup_{t \in T_{out}(S_2)} W(t, S_2) \cap V(S_1) = \emptyset \end{cases} \tag{5}$$

Now we can state that if the conditions (3), (4) and (5) are satisfied, then the services S_1 and S_2 don't interact. We call such a services **independent**. For instance, Terminate Call Screening and Reverse Charge services are independent. Note that a service is not independent from itself.

What happens now if we try to add a service S_3 to the system that contains already two independent services S_1 and S_2. It is easy to see that if S_3 is independent with respect to both S_1 and S_2, then the composition of the three services is non-interacting.

Inherited Interactions The following systems are true for the composition $S_1 \oplus S_2$:

$$\begin{cases} T_{in}(S_1 \oplus S_2) = T_{in}(S_1) \cup T_{in}(S_2) \\ T_{out}(S_1 \oplus S_2) = T_{out}(S_1) \cup T_{out}(S_2) \\ J_{in}(S_1 \oplus S_2) = J_{in}(S_1) \cup J_{in}(S_2) \\ J_{out}(S_1 \oplus S_2) = J_{out}(S_1) \cup J_{out}(S_2) \\ V(S_1 \oplus S_2) = V(S_1) \cup V(S_2) \end{cases} \tag{6}$$

Moreover, $\forall t \in T_{out}(S_1 \oplus S_2).\ W(t, S_1 \oplus S_2) = W(t, S_1) \cup W(t, S_2)$.

Let us add a service S_3 to the composition of independent services $S_1 \oplus S_2$. Suppose that S_2 and S_3 are independent but S_1 and S_3 aren't.

$$\begin{aligned} T_{in}(S_1 \oplus S_3) \cap T_{in}(S_2) &= (T_{in}(S_1) \cup T_{in}(S_3)) \cap T_{in}(S_2) \\ &= (T_{in}(S_1) \cap T_{in}(S_2)) \cup (T_{in}(S_3) \cap T_{in}(S_2)) \quad (7) \\ &= \emptyset \end{aligned}$$

thanks to the independence of $S_1 - S_2$ and of $S_2 - S_3$. For the same reason:

$$T_{in}(S_1 \oplus S_3) \cap T_{out}(S_2) = \emptyset \tag{8}$$

$$T_{out}(S_1 \oplus S_3) \cap T_{in}(S_2) = \emptyset \tag{9}$$

$$J_{in}(S_1 \oplus S_3) \cap T_{out}(S_2) = \emptyset \tag{10}$$

et cætera for the rest of equations.

As for variables, we note that

$$\begin{aligned} \bigcup_{t \in T_{in}(S_1 \oplus S_3)} W(t, S_1 \oplus S_3) \cap V(S_2) &= \\ \bigcup_{t \in T_{in}(S_1 \oplus S_3)} [W(t, S_1) \cup W(t, S_3)] \cap V(S_2) \end{aligned} \tag{11}$$

This intersection is equivalent to the union of two sets:

$$\bigcup_{t \in T_{in}(S_1 \oplus S_3)} W(t, S_1) \cap V(S_2) \ \cup \ \bigcup_{t \in T_{in}(S_1 \oplus S_3)} W(t, S_3) \cap V(S_2)$$

The first one is equal to $\bigcup_{t \in T_{in}(S_1)} W(t, S_1) \cap V(S_2)$ because the service S_1 processes only tokens from $T_{in}(S_1)$. This set is empty since S_1 and S_2 are independent (see (5)). For the same reason,

$$\bigcup_{t \in T_{in}(S_1 \oplus S_3)} W(t, S_3) \cap V(S_2) \ = \ \bigcup_{t \in T_{in}(S_3)} W(t, S_3) \cap V(S_2) \ = \ \emptyset$$

since S_3 and S_2 are independent.

Once again, because of independence between S_1 and S_3:

$$\bigcup_{t \in T_{in}(S_2)} W(t, S_2) \cap V(S_1 \oplus S_3) \ =$$
$$\bigcup_{t \in T_{in}(S_2)} W(t, S_2) \cap V(S_1) \ \cup \ \bigcup_{t \in T_{in}(S_2)} W(t, S_2) \cap V(S_3) \ = \ \emptyset$$

We can see now that the composition $S_1 \oplus S_3$ is and the service S_2 are independent. That means that the composition $S_1 \oplus S_2 \oplus S_3$ presents *the same interactions* as $S_1 \oplus S_3$. We say then that $S_1 \oplus S_2 \oplus S_3$ **inherits** the interactions from $S_1 \oplus S_3$. For instance, the composition Call Forwarding – Call Forwarding – Voice Mail presents the same interaction as the composition of two Call Forwardings (a possible looping).

3.2   The Number of Combinations to Test

Without service independence, one must test all combinations of services. For most services there is a maximum number of instances beyond which no new interactions can occur. Below we suppose that we have already tested the combinations of n instances of the service S with the rest of services in the system.

Mono-stack Services Some services only send tokens to their own stack (for example OCS, TeenLine). For such a service S:

$$\begin{cases} S \ \in \ serv(A) \\ dest(T_{out}(S)) \ = \ A \end{cases} \tag{12}$$

We call S **mono-stack** service. Since it doesn't send tokens to any other stacks, interactions can only occur in two cases: either S violates the invariants of the services located in its own stack, or exterior services send tokens that violate the invariant of S. All these combinations have already been tested after adding the first instance of S. So, new instances of a mono-stack service don't create any new interactions.

One-destination Services Many services are one-destination, that is, they only send tokens to one stack other than theirs (for example, payment services). Let us see in which situations an interaction can occur. In addition to the cases described for mono-stack services, a one-destination service S can violate the invariant of the target stack. Meanwhile, all these combinations have already been tested with one instance of S, except one – when an instance of S sends tokens to another instance of S. After that – no new combinations can be found and thus – no new interactions. The number of instances to test is thus equal to 2.

Warning. Some services, like Call Forwarding, can be parameterized in such a way that adding new instances can still create untested cases. Suppose, for instance, that we have tested the system with CF forwarding calls from A to B. A new instance can add forwarding calls from B to C, then – from C to D etc – with every new instance, one can create a longer chain of forwarded calls (and possibly a longer loop). Nevertheless, even if newer combinations are topologically different from older ones, the nature of the interaction is the same (in the case of CF – a loop).

Multi-destination Services A service can send tokens to many stacks at a time. For instance, Three Way Calling service sends tokens to itself, to SYS and to two other participants in a three-way call. After testing the first instance of a multi-destination service S, many untested combinations remain – precisely those where S sends tokens to other instances of S. Generally, if S sends tokens to n other stacks, then one must test the system up to $n+1$ instances of S, i.e., 2^n different combinations. Sometimes the number of cases can be reduced – for example SYS cannot subscribe to several services, and respectively, ordinary stacks cannot subscribe to some system services. S. Reiff in his recent paper [17] discusses in more details the problem of number of tests in services combinations.

3.3   Detection of Feature Interactions

We use model-checking to detect feature interactions. The model-checker is based on an animator tool written in LISP. The animator can simulate call scenairios written in the following form:

```
A.offhook-A.dial(B)-B.offhook-A.onhook-B.onhook
```

It simulates token creation and processing following the SSM operational semantics rules. The result of an animation is a set of states of the system. The model-checker builds an execution tree starting from the initial state (all phones are idle) and runs all the scenarios using the animator and checking services' properties preservation. Its performance is quite low, but we didn't focus on improving it since the aim wasn't an industrial application.

We tested twelve services from the Contest and have detected 24 two-way interactions and 215 three-way ones. The results are published by the Contest committee in [16].

Quite often, an interaction manifests itself by violating the invariant of POTS. This invariant is a collection of properties over variable values and translates some "common sense" requirements, for example *"If a telephone is off-hooked, then it is busy"* or *"If a telephone is communicating then it is off-hooked"* etc.

Model-checking is also useful for validating services. The validation consists of checking whether the service doesn't interact with POTS. Many specification errors can thus be corrected.

A special method is introduced to detect loops. Loops cannot be detected simply by testing safety properties – one needs also liveness ones. In our model-checker, we make a simplification – we assume that tokens pass instantly between stacks and that user actions don't occur in the same moment. That makes token exchange be cyclic – a user releases a token – the system processes it and before the processing is done, no other user releases tokens. A loop occurs when the system takes an infinite time to process a token. This happens when at some moment between two user actions the state of the system returns to an already visited state. Our model-checker tests this by memorizing all visited states between two user actions. Here are the cases where loops have been detected: Call Forward and Call Forward, Call Forward and Group Ringing, Group Ringing and Group Ringing.

4 Towards a Feature Interaction Manager

4.1 Resolution of Interactions

Because of the structure of our model, services in the same stack are ordered by priority. This already avoids some interactions, mentioned and used by others before [4, 6, 9]. Not all interactions can be avoided this way. In some cases, the disposition of services in a stack is not sufficient to prevent interactions. For example, even if the classic case of the interaction between Three Way Calling and Call Waiting (example 2 in [7]) is avoided (because CW>TWC), another situation leads to an odd behavior[1] and a variable should be introduced to control to which service the "flash" button is currently bound. Furthermore, the stack structure does not help when an interaction appears between different stacks.

Model-checking service combinations helps not only to detect interactions, but also to tell in which situations and which tokens cause an interaction.

---

[1] A is subscriber to TWC and CW. A talks to C and decides to make a new call with TWC, so he pushes "flash". At this moment B calls A. A changes his mind and wants to return to C, but pushing "flash" again connects him to B since CW>TWC.

This allows us to install a **supervisor** that knows all critical situations and controls the token flow. Once it recognizes that the situation is critical, it can remove, add and change tokens being responsible of the situation. For example, the supervisor can be notified about creation of a Call Forwarding loop. It decides then to retrieve the token **alert** that bounds the loop and to send, say, a **busy** token to the caller.

The supervisor is specified in SSM just like any service. The difference is that it has no own variables and no invariant. The supervisor filters no longer use the keyword THIS since every token should be tested regardless of its origin or destination.

Here is an example how the interaction between Call Transfer (CT) and Call Forwarding (CF) services is resolved. Call Transfer service allows the subscriber to bring two phones into a conversation (it is mostly like an operator call). Suppose A a CT subscriber. B calls A and asks him to call C. C is a CF subscriber that forwards his calls on D. After dialling the number of C, A will finally be joined to D and connect B to D. Since the connection is established in the moment when A goes onhook, B will start to pay for the line BD. Nevertheless, B called initially C (via A) and according to the CF specification, B should pay for the part BC and C – for the part CD of the line. The interaction is resolved by testing the presence of **billing_forward** token sent by CF.

```
GET billing_forward(C, SYS, A, D){
    IF (CT ∈ Serv(A)) ∧ (A.ct_third ≠ "?") THEN
        SEND billing_forward(C, SYS, A.other, D) UP
    END
    SEND billing_forward(C, SYS, A, D) UP
    }
```

In our example, $A.other$ refers to B, so the supervisor makes the billing machine think that the forwarding call comes from B. Some animations and model-cheking show that the interaction disappears.

We used the supervisor to resolve the most of the earlier detected interactions. The interactions that have not been resolved are all take place because the services specifications are contradictory. In these cases a choice has to be made to accord a priority to one of the interacting services.

This supervisor method is similar to the method of N. Fritsche [10]. The supervisor has a disadvantage—with every new interaction resolution one needs to re-test the system to be sure that the resolution hasn't created new interactions. In fact, since the supervisor enters in action only on particular discrete situations, it can rarely create interactions.

4.2   Abstract Specifications

In order to improve the communication with the customer, we propose to introduce an abstract specification method. The idea is to represent some token exchanges by macros—easy to understand and to combine. The abstract level can be specified in several ways. We are actually testing a graphical formalism—state Chisel diagrams [1].

In state Chisel diagrams, nodes are states and transitions are actions. In SSM, a state of a stack corresponds to a collection of tokens in the stack and the set of stack variables together with their values. An action is the sending of a token. We can make Chisel diagram nodes correspond to states of a stack. Chisel transitions will indicate which tokens lead to what state changes. Having all this information, it becomes quite easy to represent Chisel diagram in SSM. We show this using a simple example.

Let us start with POTS. To make a call, the user takes off the receiver and dials the number. The Chisel diagram of making a call is the following

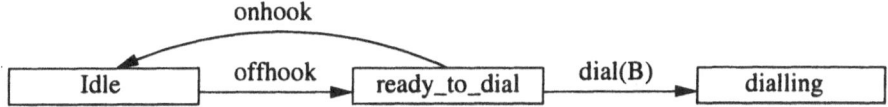

To translate it into SSM we do as follows:

1. Choose the set of stack variables. A good choice is a matter of experience. All the states should be captured by variable/value combinations. For POTS we propose the following variables:

   VARIABLES

busy,	true if the telephone is not idle
talking,	true if the telephone is communicating
rdown,	true if the receiver is down. (DOWN is a keyword in SSM, that is why the variable is called *rdown*)
other,	the other's telephone (if known)
tone	the tone — "silent" , "ringing" etc

2. Code in SSM the nodes of the diagram. For example, the node "Idle" is represented by the combination

   busy=false; *talking*=false; *rdown*=true; *other*="?"; *tone*="silent"

   The state "ready_to_dial" corresponds to the combination

   busy=true; *talking*=false; *rdown*=false; *other*="?"; *tone*="dial_tone"

3. We know now that the action of sending an **offhook** token makes the stack pass from the state "Idle" to the state "ready_to_dial". Therefore, we can make a filter accepting tokens of type **offhook** that changes states as appropriate:

GET **offhook**(THIS, THIS){
  $busy$:= true
  $rdown$:= false
  $tone$:= "dial_tone"
  }

4. Later, this filter will need to be changed, since there are other offhook actions. Some conditions will be introduced (like testing whether the stack is in the right state) to define an appropriate processing of the **offhook** token.

5. Repeat the steps 2–4 for all diagram.

Safety properties cannot be expressed in terms of diagrams, they have to be added when the translation is done. An animation can be useful to check whether the SSM specification allows to follow all the traces of the Chisel diagram. Service specifications can be written in a short form (like those of the First Feature Interaction Detection Contest [12]) – omitting common parts with POTS. Having common parts with POTS means in SSM that POTS processes tokens corresponding to them and the service doesn't need to do anything with them.

The choice of Chisel is not mandatory. We suppose that other specification methods can be used on the abstract level. We have tested the Description Logic of A. Gammelgaard and J.E. Kristensen [11]; the results are satisfying. The use of other powerful formal methods on the abstract level allows to detect many interactions before passing to the concrete level of SSM. The role of SSM is then to define an executable model and to detect the rest of interactions – either those that have not been detected by the previous methods, or those that still don't appears on the abstract level.

5 Conclusion

SSM is not only an architecture model. It presents a complete system of service development and verification allowing a simple communication with the customer by means of diagrams. It is reinforced by an animation tool and a model-checker that helps to detect eventual feature interactions. It offers also a technique for interactions resolution – by using a token supervisor.

Some points need improvement or further development.

- SSM is a particular case of coordination languages [8]. It would be intresting to apply some results of coordination model theory to see whether interactions can be detected by proof, not by model-checking.

- The collaboration with other formal methods (on the abstract level) should be studied. A translation to SSM should be figured out for every method.
- Even if the U-door allows to cope with some eventual network problems, it would be intresting to study the impact of network anomalies on our model.
- It is tempting to apply SSM techniques to other systems than the telecommunications world – Internet, communicating agents, operating systems etc.

References

1. A. Aho, S. Gallagher, N. Griffeth, C. Schell, and D. Swayne. SCF3™/Sculptor with Chisel: requirements engineering for communications services. In K. Kimbler and L.G. Bouma, editors, *Feature Interactions in Telecommunications Systems V*, pages 45–63. IOS Press, Amsterdam, September 1998.
2. J.M. Atlee and K.H. Braithwaite. Towards automated detection of feature interactions. In W. Bouma and H. Velthuijsen, editors, *Feature Interactions in Telecommunications Systems*, pages 36–57, Amsterdam, May 1994. IOS Press.
3. P.K. Au and J.M. Atlee. Evaluation of a state-based model of feature interactions. In P.Dini, R. Boutaba, and L. Logrippo, editors, *Feature Interactions in Telecommunications and Distributed Systems IV*, pages 153–167, Amsterdam, June 1997. IOS Press.
4. M. Bostrm and M. Engstedt. Feature interaction detection and resolution in the delphi framework. In K.E. Cheng and T. Ohta, editors, *Feature Interactions in Telecommunications Systems*, volume 3, pages 157–172. IOS Press, Amsterdam, June 1995.
5. J. Bredereke. Automata-theoretic vs. property-oriented approaches for the detection of feature interactions in IN. In T. Margaria, editor, *Int'l. Workshop Advanced Intelligent Networks (AIN'96)*, pages 56–70, Passau, March 1996. Universität Passau, Fakultät für Mathematic und Informatik.
6. J. Bredereke and R. Gotzhein. A case study on specification, detection and resolution of IN feature interactions with Estelle. Technical Report 245/94, Univ. of Kaiserslautern, Dept. of Comp. Sc., May 1994.
7. E.J. Cameron, N.D. Griffeth, Y.-J. Lin, M.E. Nilson, W.K. Schnure, and H. Velthuijsen. A feature interaction benchmark for IN and beyond. In W.Bouma and H.Velthuijsen, editors, *Feature Interactions in Telecommunications Systems*, pages 1–23. IOS Press, Amsterdam, May 1994.
8. N. Carriero and D. Gelernter. Coordination languages and their significance. *Communications of the ACM*, 35(2):97–107, 1992.
9. Y.L. Chen, S. Lafortune, and F. Lin. Resolving feature interactions using modular supervisory control with priorities. In R. Boutaba P. Dini and L. Logrippo, editors, *Feature Interactions in Telecommunications Systems*, pages 108–122. IOS Press, Amsterdam, June 1997.
10. N. Fritsche. Runtime resolution of feature interactions in architecture with separated call and feature control. In K.E. Cheng and T. Ohta, editors, *Feature Interactions in Telecommunications Systems*, volume 3, pages 43–63. IOS Press, Amsterdam, June 1995.

11. A. Gammelgaard and J.E. Kristensen. Interaction detection, a logical approach. In L. Bouma and H. Velthuijsen, editors, *Feature Interactions in Telecommunications Systems*, pages 178–196. IOS Press, Amsterdam, May 1994.

12. N. Griffeth, R. Blumenthal, J.-C. Grgoire, and T. Ohta. Feature interaction detection contest. In K. Kimbler and L.G. Bouma, editors, *Feature Interactions in Telecommunications Systems V*, pages 327–359. IOS Press, Amsterdam, September 1998.

13. M. Jackson and P. Zave. Distributed feature composition. a virtual architecture for telecommunications services. In *IEEE Transactions on Software Engineering*, volume 24-10, pages 831–847. IEEE, October 1998.

14. D.O. Keck and P.J. Kuehn. The feature and service interaction problem in telecommunications systems: A survey. *IEEE Transactions on Software Engineering*, 24(10):779–796, October 1998.

15. K. Kimbler. Service interaction in next generation networks: Challenges and opportunities. In M. Calder and E. Magill, editors, *Feature Interactions in Telecommunications and Software Systems VI*, pages 14–20. IOS Press, Glasgow, May 2000.

16. M. Kolberg, E.H. Magill, D. Marples, and S. Reiff. Second feature interaction contest. In M. Calder and E. Magill, editors, *Feature Interactions in Telecommunications and Software Systems*, pages 300–325. IOS Press, Glasgow, May 2000.

17. S. Reiff. Notes on call configurations with features. In S. Gilmore and M. Ryan, editors, *Workshop on Language Constructs for Describing Features*, pages 71–77, Glasgow, May 2000.

18. P. Zave and M. Jackson. New feature interactions in mobile and multimedia telecommunication services. In M. Calder and E. Magill, editors, *Feature Interactions in Telecommunications and Software Systems VI*, pages 51–66. IOS Press, Glasgow, May 2000.

The Declarative Language STR (State Transition Rule)

T. Yoneda and T. Ohta

SOKA University, Faculty of Engineering, 1-236, Tangi-cho, Hachioji-shi, Tokyo 192-8577 Japan. Email: anne@t.soka.ac.jp, ohta@t.soka.ac.jp

Abstract. The declarative language STR (State Transition Rule) is proposed to describe service specifications and programs. Given that telecommunication service specifications can be represented in the form of a state transition diagram, STR describes conditions for state transitions in the form of a production rule. STR specifications, description examples, categories of feature interactions, examples of feature interactions, interaction detection method, and application to Active Networks are described.

1 Introduction

This paper proposes a declarative language, STR (State Transition Rule). STR was developed at ATR(Advanced Telecommunication Research Institute) originally to describe telecommunication service specifications for automatically generating switching systems programs. STR has also been used to describe service specifications to automatically detect feature interactions. Now, it is also used as programming language for prototyping systems.

In this paper, the birth of STR, its specifications as a telecommunication service specification description language, its application in detecting feature interactions and its recent application to Active Networks are described.

Considering that telecommunication service specifications can be represented in the form of a state transition diagram, STR was developed to describe conditions for state transitions as production rules. It exhibits a number of characteristics. Among them, because of application rules for production rules, feature interactions between service A and service B occur solely by combining both sets of rules for service A and service B. This leads formal methods for detecting feature interactions without state creations. Another characteristic is that by using STR as a programming language it is very easy to add new services to existing services.

In section 2, the original objective for developing STR is described. In section 3, language specifications of STR are explained briefly. In section 4, some examples of describing services using STR are shown. In section 5, the following feature interaction issues are discussed: formal definitions of feature interactions, some examples of feature interactions, non-monotony phenomenon in adding new services, and interaction detection methods based on STR. In section 6, application to Active Networks is reported.

2   Objectives for developing STR

STR was developed at ATR(Advanced Telecommunication Research Institute) originally to describe telecommunication service specifications for automatically generating switching systems programs [1] (Figure 1). Potential users are those who can understand telecommunication services but are not necessarily experts on switching systems. In the Interface Process block,

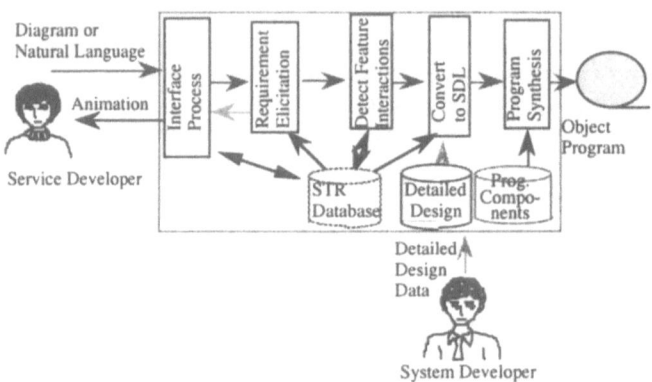

Fig. 1. Automatic programming system

specifications using diagram form or natural language are converted to STR. Also, animation for service specification input by a user is shown to a user to confirm that her or his description has no mistakes. In the Requirement Elicitation block, the set of rules described using STR for input specification is validated as a single service. In the Detect Feature Interactions block, STR for input service specification is validated if it causes feature interactions with other services specifications. In the Convert to SDL block, all rules described in STR are merged and converted to SDL (Service Description Language), which is one of standard service description languages. In this process, a detailed description of SDL, which has been stored beforehand by system developers, is added automatically. In the Program Synthesis block, the specification described in SDL is converted to a C language program automatically. In this process, program components, which have been stored beforehand by system developers, are added.

Thus, in so far as input specification requires only prestored precise SDL descriptions and program components, input service specification is automatically converted to C program.

qSTR has also been used to describe service specifications to automatically detect feature interactions [2][3][4][5][6]. Now, it is also being used as a programming language for prototyping systems.

As is well known, telecommunication service specifications can be described in the form of a state transition diagram. Therefore, the service specification can be represented as a set of rules which describe conditions of the state transitions.

3   Syntax and application rules

A system state is represented as a set of statuses, called primitives, of terminals or relationships between terminals connected to the system. The primitive consists of a primitive name, which represents a status, and arguments which indicate real terminals (e.g. A,B,C).

The primitives which begin with m- represent a service being activated. For example, if terminal A has Call Waiting service activated it is represented as m-cw(A). The primitives which begin with m- do not necessarily change during a call s state transition from initial state, idle(A), to final state, idle(A) in the case of telecommunication services.

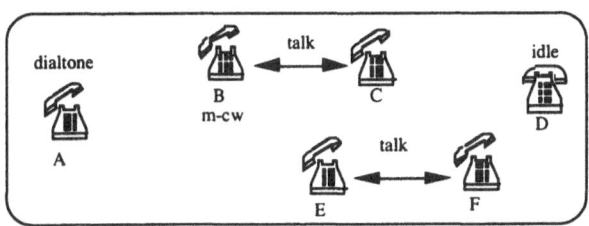

Fig. 2. System state

Example 1: Terminal B has Call Waiting service activated (denoted as m-cw(B)), terminal A is hearing a dial-tone (denoted as dialtone(A)), terminal B and C are talking with each other (denoted as talk(B,C)), terminal D is in idle state (denoted as idle(D)), terminal E and F are talking with each other (Figure 2). Then, this system state is described as follows:
   {m-cw(B),dialtone(A),talk(B,C),idle(D),talk(E,F)}
The syntax of STR is as follows.
Pre-condition event: Post-condition
Pre-condition and Post-condition are represented as a set of primitives, respectively. In the rule, the arguments of each primitive are described as terminal variables (e.g. x,y,z) so that the rule can be applied to any terminals.

Example 2: Suppose, Call Waiting service. Terminal y has CW service activated. Terminal x is hearing a dial-tone, and terminal y and z are talking with each other. When terminal x dials to terminal y (denoted as dial(x,y)), terminal y transits to call waiting ringing state (denoted as cw-ringing(y,x)) and terminal x hears audible ringing tone (denoted as cw-ringback(x,y)). In this case, the rule is described as follows:

m-cw(y),dialtone(x),talk(y,z) dial(x,y):
m-cw(y),cw-ringing(y,x),cw-ringback(x,y),talk(y,z)

Next, rule application and change of system state when the rule is applied are described. The states where for Pre-condition or Post-condition terminal variables are replaced by the real terminals are called the state corresponding to Pre-condition or Post-condition, respectively. When the state corresponding to Pre-condition exists in the system state, it is said that Pre-condition is included in the system state.

Rule 1) Basic rule application: When an event occurs in a system, a rule that has the same event and whose Pre-condition is included in the system state is applicable.

Rule 2) Precedent rule application: When more than one rule is applicable, the rule whose Pre-condition includes any other rules' Pre-conditions is applied.

Rule 3) Change of system state: When a rule is applied, the next state of the system state is obtained as follows: The state corresponding to Pre-condition of the applied rule is deleted from the current system state and the state corresponding to Post-condition of the applied rule is added.

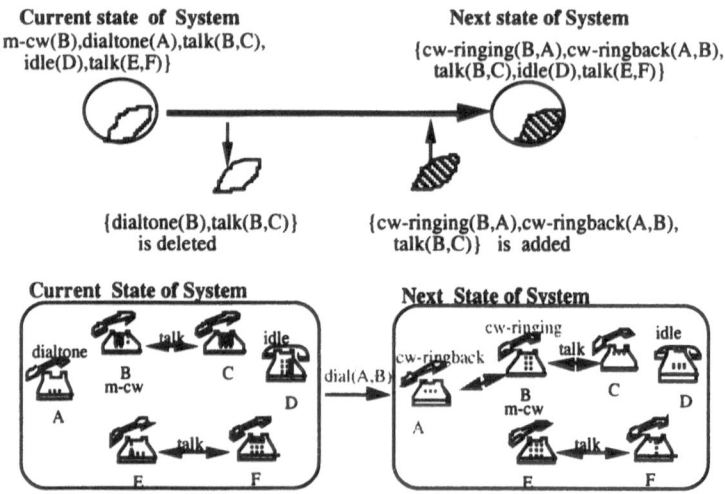

Fig. 3. System state change

Example 3: Let the current system state be the same as that described in example 1, and the rule, which is the same as described in Example 2, is applied. Suppose an event, dial(A,B), occurs in the system. Terminal variables x, y and z in Pre-condition of the rule can be replaced by real terminals A, B and C. Then, the state corresponding to the Pre-condition of the rule is obtained as {m-cw(B),dialtone(A),talk(B,C)}. Since the Pre-condition of the rule is included in the system state, the rule can be applied. When the rule is applied the state corresponding to the Pre-condition of the rule is deleted from the system state, and the state corresponding to the Post-condition is added (Figure 3).

4   Examples for STR descriptions

4.1   Basic service

STR description for basic service, normal route, is shown in Figure 4.

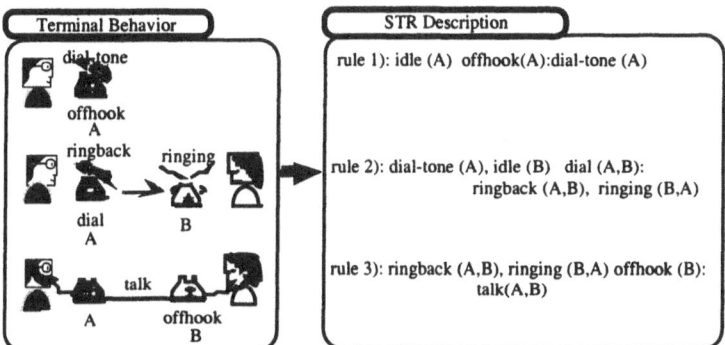

Fig. 4. STR description for basic service

4.2   Call forwarding service

As STR descriptions for basic service are automatically used for Call forwarding service, STR descriptions to be added for Call forwarding service, normal route only, are shown as follows:

rule cf1): dialtone(x),m-cfv(y,z),idle(z) dial(x,y):

ringback(x,z),ringing(z,x),m-cfv(y,z)

rule cf2): dialtone(x),m-cfv(y,z),not[idle(z)] dial(x,y): busy(x),m-cfv(y,z)

4.3   Call waiting service

In the same way, STR descriptions to be added for Call waiting service, normal route only, are shown as follows:

rule cw1): m-cw(x),talk(x,y),dialtone(z) dial(z,x):
m-cw(x),talk(x,y),cw-ringback(z,x),cw-ringing(x,z)
rule cw2): m-cw(x),talk(x,y),cw-ringback(z,x),cw-ringing(x,z) flash(x):
m-cw(x),hold(x,y),talk(x,z)
rule cw3): m-cw(x),hold(x,y),talk(x,z) flash(x): m-cw(x),hold(x,z),talk(x,y)
rule cw4): m-cw(x),hold(x,y) onhook(y): m-cw(x),idle(y)
rule cw5): m-cw(x),hold(x,y),talk(x,z) onhook(z): m-cw(x),talk(x,y)
rule cw6): m-cw(x),talk(x,y),hold(x,z) onhook(x):
m-cw(x),busy(y),ringing(x,z),hold(x,z)
rule cw7): m-cw(x),ringing(x,z),hold(x,z) offhook(x): m-cw(x),talk(x,z)

4.4   Originating call screening service

For originating call screening service, if an originating terminal dials to a terminal which has been registered in the screening list as a screened terminal, the call connection is rejected (Figure 5). Not only the directory number for an individual terminal but also the special code, in order to inhibit distance call, can be registered in the screening list. To add the originating call screening

Fig. 5. Originating call screening service

service, for a normal route, only the following rule has to be added:
dialtone(x),m-ocs(x,y) dial(x,y): m-ocs(x,y),busy(x)

4.5   Terminating Call Screening Service

For terminating call screening service, if a call terminated from a terminal which has been registered in the screening list as a screened terminal, the call connection is rejected (Figure 6). Both the directory number for individual terminals and area codes for rejecting calls from certain areas, can be registered in the screening list. To add the terminating call screening service, for normal route, only the following rule has to be added:
m-tcs(y,x),dialtone(x) dial(x,y): m-ocs(y,x),busy(x)

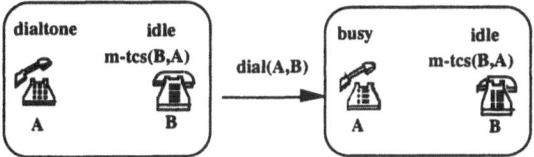

Fig. 6. Terminating call screening service

4.6 Feature Integration

When a feature is integrated into existing features, the union of the rules forming existing features and the rules forming the feature is taken. To detect feature interactions between two features, the union of the rules forming each service is taken.

5 Feature Interactions

5.1 Formal Definitions

To detect feature interactions automatically[7]-[11], formal definitions of feature interactions are needed. Given that telecommunications service specifications can be written as a form of a state transition diagram and, therefore, feature interactions can be represented as abnormal state transitions, author has proposed formal definitions of feature interactions for the following seven categories [4].

- dead lock
- live lock
- non-determinacy
- appearance of an abnormal state
- disappearance of a normal state
- appearance of an abnormal transition
- disappearance of a normal transition

Dead lock, live lock, and non-determinacy can be identified from the diagram. For the four other interactions, we cannot identify from whether or not interactions occur without knowing the meaning of transitions and states. Therefore, the first three interactions are called logical interactions, and the remaining four interactions are called semantic interactions.

5.2 Non-determinacy

Non-determinacy is defined in such a way that for an event more than one transition is possible. In other words, for an event the sate transition cannot

be determined. Suppose, terminal A has call waiting service and call forwarding service activated, and has registered terminal D as a forwarded terminal.
When terminal C dials to terminal A while terminal A and terminal B are
talking with each other, the system cannot decide which service should be
applied, call waiting service or call forwarding service (Figure 7).

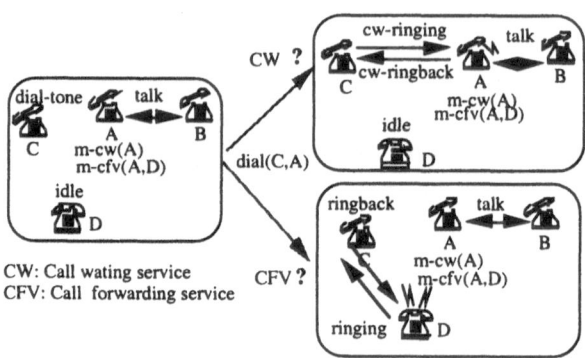

Fig. 7. Non-determinacy

5.3   Other examples

According to formal definitions on feature interactions, some feature interactions between call waiting service and call forwarding service are shown.
As described in section 5.2, if terminal D dials B when terminal A and terminal B are talking with each other and terminal B has call waiting service
and call forwarding service activated, non-determinacy happens. In the same
case, suppose terminal B has registered terminal C as a forwarded terminal
to which any calls to terminal B are forwarded, and a system decides that
call forwarding service takes precedence over call waiting service. Then, in
this case, the call from terminal D to terminal B is forwarded to terminal
C. From the call waiting service's view point, two abnormal state transitions
occur. One is disappearance of a normal state, and the other is appearance
of an abnormal transition (Figure 8). If call forwarding service takes precedence over call waiting service, cw-calling state disappears. cw-calling(D,B)
state represents the state where, if the new call terminates to terminal B
while terminal B has call waiting service activated and is talking with terminal A, terminal B hears the special audible ringing tone which indicates
a new call terminates to terminal B. The state cw-calling(D,B) can be transited only from the state talk(A,B). Therefore, if call forwarding service takes
precedence over call waiting service, cw-calling(D,B) never appears.

Moreover, a transition from talk(A,B) to cw-calling(D,B) is an intention
of call waiting service. Therefore, from the call waiting service's view point,

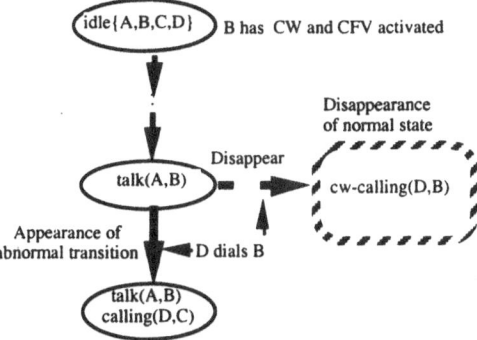

Fig. 8. Interactions in case that CFV takes precedence over CW

transition to the new state transited by call forwarding service is abnormal, appearance of an abnormal transition.

On the other hand, the system decides that call waiting service takes precedence over call forwarding service. In this case two interactions can be seen, appearance of abnormal transition and disappearance of normal transition (Figure 9).

If call waiting service takes precedence over call forwarding service while terminal A and B are at the state talk(A,B), a new call to terminal B is never forwarded to terminal C. But, the state {talk(A,B),calling(D,C)} can be reached via another route as shown in Figure 9. Therefore the state {talk(A,B),calling(D,C)} does not disappear. Therefore, this is not a case for disappearance of a normal state but disappearance of a normal transition. Another interaction occurs in a transition from talk(A,B) to {talk(A,B),cw-

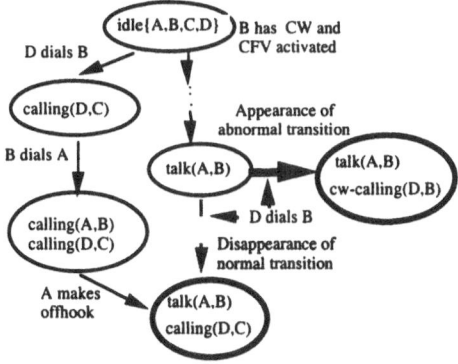

Fig. 9. Interactions in case that CW takes precedence over CFV

calling(D,E)} in Figure 9. From the call forwarding service's view point, the transition from talk(A,B) to {talk(A,B),calling(D,C)} is an intention. But, in a transition from talk(A,B) to {talk(A,B),cw-calling(D,E)}, the intention is not realized. Therefore, from the call forwarding service's view point, the transition from talk(A,B) to {talk(A,B),calling(D,C)} is abnormal.

5.4   Non-monotony

As shown in the previous section, 5.3, when adding new services to existing services, some states or/and some transitions may disappear. This shows non-monotony phenomenon in adding a new service. We can say that feature interactions cause a non-monotony phenomenon in adding new services.

5.5   Interaction detection method based on STR

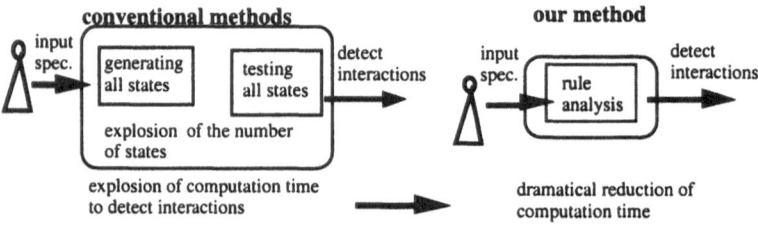

Fig. 10. Comparison of detecting methods

We have proposed formal methods for detecting feature interactions under the condition where service specifications are described using STR[5][6]. They deals with semantic interactions. Though they are now under evaluation, the proposed methods are at present efficient and effective. Since the proposed methods do not require the generation of any states, they do not cause an explosion of the number of states no matter how many terminals are connected to the network. This results in a dramatic reduction of computation time for detecting feature interactions. See Figure 10.

A brief explanation of our method is given. For detailed descriptions, please refer to paper [5] and [6].

Semantic interactions can be considered as follows: Suppose two services are activated. When either specification of the services is applied, a state transition according to the specification contradicts the specification for the other service. Feature interactions are then detected as follows: From each service, select a rule, respectively, which is applicable to the same system state. Apply either rule to the system state. Check if the state transition by the rule causes abnormal state transition from the view point of the other service, whose rule is not applied.

According to conventional detection methods, all possible states must be generated by one way or another and all state transitions should be checked to detect feature interactions. This causes an explosion of the number of states causing an explosion of computation time to detect feature interactions.

According to our method, however, interactions are detected solely by analyzing Pre-conditions, events, and Post-conditions of selected rules. This method does not require any state creation and does not cause an explosion of computation time for detection of feature interactions.

6 Application to Active Networks

We used STR to describe a program up-loaded to VoIP gateway which adopts Active Networks architecture[12][13]. The following are experimental results.

6.1 ESTR

ESTR was developed as a programming language for users to describe their programs which are up-loaded to Active Networks. It was developed by enhancing STR. More precisely; as a condition for rule application, conditions for state transitions are applied in the same way as STR, and a description part of conditions for system controls required for the state transition is added.

ESTR has the form of Pre-condition, event and Post-condition. It is a rule for defining a condition for state transition, state change while the rule is applied, and system control required for the state transition. Pre-condition consists of status description elements called primitives. Primitives are statuses of terminals or relationships between terminals which are targets of the state transition. An event is a trigger which causes the state transition, e.g. a signal input to the node and some trigger occurs in the node. Post-condition consists of two parts. One is the state description part which also consists of primitives. The other is the system control description part which indicates the system controls required for the state transition. The system control description part is described in {} which follows after state description part separated by ', '(see Figure 11). When no system controls are required, the content of {} is empty. An example of ESTR is shown in Figure 11. The

call(x,y) connotify(y,x): talk(x,y),{Send(con,y,x),Con(x,y)}

Fig. 11. An example of ESTR

example in Figure 11 is explained. Terminal x and y are in calling state, denoted by call(x,y). If terminal y makes offhook, denoted by connotify(y,x), a signal Connect is sent to terminal x, denoted by Send(con,y,x), and terminal x and y transit to talk state, denoted by talk(x,y). call(x,y) and talk(x,y) are called status primitives. All arguments in status primitives are described as variables so that a rule can be applied to any terminals.

6.2   Experimental System

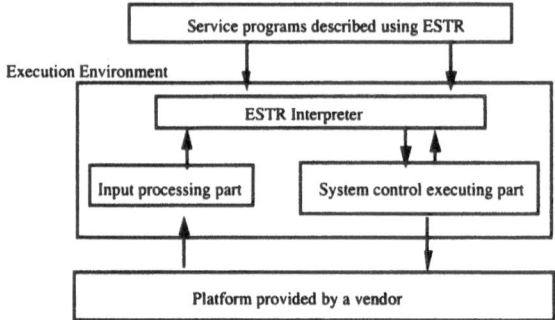

Fig. 12. Experimental system

The software structure of an experimental system for the proposed system is shown in Figure 12. An execution environment program, which consists of

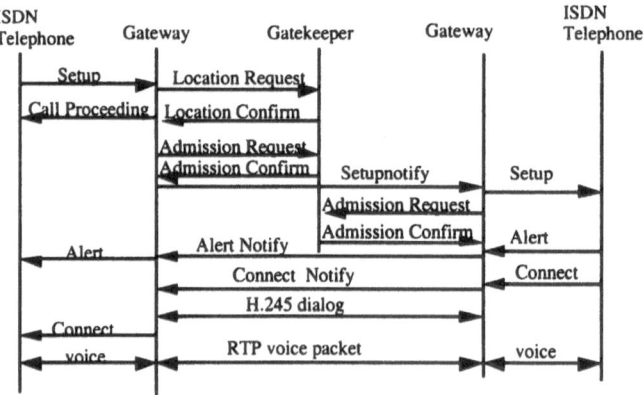

Fig. 13. H.323 Protocol

an Input processing part, an ESTR Interpreter and a System control executing part, is implemented on top of an IP Gateway Platform purchased on the market. A user program described using ESTR is executed on the Execution environment program. The protocol between terminals and the IP Gateway is ISDN. The system control description part shows conditions for controlling the IP Gateway based on protocol H.323[14]. H.323 protocol is shown in Figure 13.

ESTR Interpreter The ESTR Interpreter is initiated by receiving an event from the Input processing part, the interpreter selects a rule in a rule data base, and interprets the rule. The interpreter sends the system control description part of the rule to a system control executing part. When the interpreter receives execution results from the system control executing part, a state corresponding to the Pre-condition of the rule is deleted from the system state, and a state corresponding to the Post-condition of the rule is added to the system state.

Input processing part When the Input processing part receives a signal from the Platform, the Input processing part translates it to an event defined in ESTR Interpreter, and the event is sent to the Interpreter to initiate the Interpreter.

System control executing part When the System control executing part receives a signal (system control description part of the rule) from the Interpreter, it analyzes the signal and calls the appropriate API provided by the Platform to send signals to terminals or other nodes.

Basic service The ESTR description of a normal route in basic service of VoIP Gateway is shown in Figure 14. In Figure 14, wtalert(x,y) represents that terminal x is awaiting 'alert' signal from terminal y. called(x,y) represents that terminal y is called by terminal x. call(x,y) represents that terminal x is calling terminal y. talk(x,y) represents that users of terminal x and terminal y are talking with each other. wtrelcomp(x) represents that terminal x is awaiting 'release completion' signal from the network. wtrel(y) represents that terminal y is awaiting 'release' signal from the network. In the same rule, the same terminal variables represent the same terminal. Between different rules, the same terminal variables, x in rule 1 and x in rule 2, are not necessarily the same terminal. On the other hand, different terminal variables in the same rule represent different terminals. But, between different rules, different terminal variables, x in rule 1 and y in rule 2, are not necessarily different terminals.

setup(x,y), alert(y,x), disc(x,y), rel(x) and relcomp(x) are events. setup(x,y) represents receiving a 'setup' signal from terminal x to terminal y. alert(y,x) represents receiving an 'alert' signal from terminal y to terminal x. disc(x,y) represents receiving a 'disconnect' signal from terminal x to terminal y. rel(x) represents receiving a 'release' signal from terminal x. relcomp(x) represents receiving a 'release complete' signal from terminal x.

Send(s,x,y) represents sending terminal y a signal 's' from terminal x. Con(x,y) represents connecting terminal x and terminal y. Disc(x,y) represents releasing a connection between terminal x and terminal y. A signal flow

```
idle(x) setup(x,y): wtalert(x,y),{Send(calp,x),Send(setupnotify,x,y)}
idle(y) setupnotify(x,y): wtalert(y,x),{Send(setup,x,y)}
wtalert(y,x) alert(y,x): called(y,x),{Send(alertnotify,y,x)}
wtalert(x,y) alertnotify(y,x): call(x,y),{Send(alert,y,x)}
called(y,x) con(y,x): talk(y,x),{Send(connotify,y,x),Send(conack,y,x)}
call(x,y) connotify(y,x): talk(x,y),{Send(con,y,x),Con(x,y)}
talk(x,y) disc(x,y): wtrelcomp(x),{Disc(x,y),Send(rel,x),Send(discnotify,x,y)}
talk(y,x) discnotify(x,y): wtrel(y),{Send(disc,x,y)}
wtrel(x) rel(x): idle(x),{Send(relcomp,x)}
wtrelcomp(x) relcomp(x): idle(x),{}
```

Fig. 14. Examples of ESTR Descriptions

from a receiving setup signal to a sending setup signal to a Gatekeeper is shown in Figure 15.

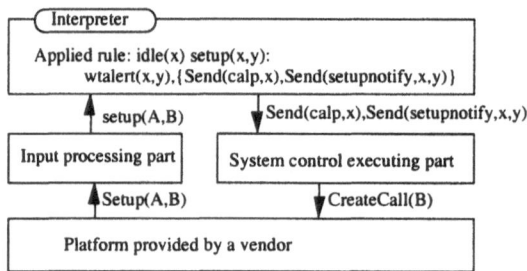

Fig. 15. An example of signal flow

7   Summary

The declarative language STR (State Transition Rule) was proposed to describe service specifications and to program a telecommunications system. Objectives and specifications of STR and some examples were described. Based on state transition model, 7 categories for feature interactions were described.

According to the categories, some examples of feature interactions were shown. An out line of our method for detecting feature interactions were described. With this method, generating system states is not required. Therefore, an explosion of computation time for the detection of feature interactions can be avoided.

An experimental system for Active Networks using STR to describe uploaded programs was shown, illustrating that by using STR as a programming language it is very easy for users to add new services to existing services.

Future work will focus on evaluation of feature interaction detection methods and their application to Active Networks, and new applications of STR.

References

1. Y. Hirakawa et al., "Telecommunication Service Description Using State Trnsition Rules", Int. Workshop on Software Specification and Design, Oct. 1991.
2. T. Ohta and et al., "Classification, Detection and Resolution of Service Interactions in Telecommunication Services", Proc. of FIW94, pp.60-72, May 1994.
3. Y. Kawarasaki and T. Ohta, "A New Proposal for Feature Interaction Detection and Elimination", Proc. of FIW95, pp.127-140, Oct. 1995.
4. T. Ohta and C. Flaviu, "Formal Definitions of Feature Interactions in Telecommunications Software", IEICE Trans. on Fundamentals, vol. E-81A, No.4, pp.635-638, April 1998.
5. T. Yoneda and T. Ohta, "A Formal Approach for Definition and Detection Feature Interactions", Proc. of FIW98, pp.202-216, Sep. 1998.
6. T. Yoneda and T. Ohta, "Automatic Elicitation of Knowledge for Detecting Feature Interactions in Telecommunication Services", IEICE Trans. on Information and Systems, vol. E-83D, No.4, April 2000.
7. L. G. Bouma and H. Velthuijsen (eds.), "Feature Interactions In Telecommunication Networks II", IOS Press, 1994.
8. K. E. Cheng and T. Ohta (eds.), "Feature Interactions In Telecommunication Networks III", IOS Press, 1995.
9. P. Dini et al. (eds.), "Feature Interactions In Telecommunication Networks IV", IOS Press, 1997.
10. K. Kimbler and L. G. Bouma (eds.), "Feature Interactions In Telecommunications And Software Systems V", IOS Press, 1998.
11. M. Calder and E. Magill (eds.), "Feature Interactions in Telecommunications and Software Systems VI", IOS Press, 2000.
12. K. L. Calvert, et al., "Directions in Active Networks", IEEE Com. Magazine, Vol.36 No.10, pp.72-78, Oct. 1998.
13. "Active Networks", Lecture Notes in Computer Science 1653, Springer, 1999.
14. ITU-T Rec. H.323, "Packet-based multimedia communications systems," 1998.

Modular Feature Integration and Validation in a Synchronous Context

Nicolas Zuanon* **

Laboratoire LSR-IMAG, BP 72,
F-38402 Saint Martin d'Hères Cedex, France

Abstract. This paper describes a methodology for specifying and validating feature-based systems. Our solution allows the easy integration of a new feature into an existing system and to perform its validation in an incremental way. Our specification proposal is based on the synchronous approach, which proved to be well-adapted to the chosen level of abstraction.

The validation process focuses on the detection of interactions between features. The validation is based on an automated testing technique driven by behavioral patterns.

We provide in this paper a view of the methodology and the operations it requires. This approach can be applied to any system whose features can be modeled by automata and implemented in a synchronous style. We illustrate here its application to the problem of telecommunication feature interaction detection.

1   Introduction

A feature is an increment to a system and it usually provides a supplementary and self-contained functionality. Systems based on features are developed by successive feature integrations, using an incremental method. As the features may be designed and developed independently, this raises the feature interaction problem.

A feature interaction occurs when several features, developed in an independent manner, are provided simultaneously by a single system. This coexistence may lead to the unexpected modification of one or several features' behaviours. This problem hampers the rapid creation and integration of new features, and increases the need for a dedicated and adapted method for the validation of features.

Validating each feature in an isolated manner does not suffice to ensure that the system which will provide them to the users will operate correctly. One must also confront the features one to the other to assert the global correction of the system.

---

* This work has been partially supported by a contract between CNET-France Telecom and University Joseph Fourier, #957B043.
** Current address: GIE DYADE, INRIA Rhone-Alpes, 655 avenue de l'Europe, Montbonnot F-38334 Saint Ismier Cedex.

Prior to the validation itself, a modeling activity is required, so to make the features intelligible in some formalisms usable for validation purposes. In this article, we chose to tackle the modeling and the validation problems in a unified manner. Both tasks are conducted in an incremental way and are strongly connected.

We more precisely propose to tackle the problems of feature validation and interaction detection at the specification level, in order to focus on the logical aspects of the features and to concentrate on the detection of misconceptions of the features. Specification handling allows to reason on abstract and simplified models, which are more easily handled. It is also a means to intervene very early in the cycle of development of the feature and to minimize the cost of an error: the sooner the error is revealed, the lower is the cost to correct it.

Modeling a specification is always a useful task, whether an implementation is automatically derived from it or not. In either case, the validation of this model allows to ensure that the specification is non-ambiguous and consistent. At the same time, it suppresses conception errors and reduces the complexity of the implementation task [11,8].

Our work is original by the following aspects:

- It distinguishes the features from the basic system and defines clearly their relation, thanks to two dedicated operations: *composition* of features and *integration* of a feature to the basic system.
- It provides a homogeneous way to handle all three specifications that compose a feature, namely the feature behavior, the feature requirements, and the feature usages. A unified framework allows to both model the features and validate them.

Modeling Most pieces of work dealing with the problem of feature modeling propose an integration operation which alters deeply the initial model[1].

As a consequence, the complexity of the model resulting from an integration is higher than the added complexities of the initial model and of the feature to be integrated. The operation of removing a feature from a model may as well become impracticable. This problem is known as non-monotonicity [16] and appears clearly as an obstacle to the evolution of such a model.

Our work shares some analogies with those of Plath and Ryan [14,15], which propose a general approach to construct in the most independent manner each feature, in order to ease afterwards their composition. The definition

---

[1] For instance, Hall [8] proposed to describe a feature by rewriting and modifying the whole basic service. Because of the numerous so-called "spurious" interactions revealed by his method, he came to the conclusion that the definition of a feature shall be restricted to the minimal modifications that it indu es on the basic service.

and the use of a dedicated operator for the feature's description allows to clearly identify the relation between the features and the basic service.

To reduce the complexity of the model, we take advantage of the synchronous approach. This approach allows to simplify the model and down-size it, thus making easier its handling. The model we propose is also executable, which allows its animation and facilitate its analysis.

The main objective of our modeling approach is to provide a highly modular specification which keeps apart the basic service and the additional features and allows to ignore this problem. A modular specification allows both the modeling and the validation of features in an incremental manner.

Validation The validation is performed using an automated testing method adapted to the problem. Interaction detection and testing are similar since their objective consists in defect searching. This is different from verification approaches based on theorem-proving or model-checking whose aim is to guarantee software correctness. Testing appears therefore to be well adapted to tackle the interaction detection problem, even though it offers a lower degree of confidence than verification.

In addition, the kind of testing we rely on is dynamic: sequences of inputs are generated "on-the-fly", based on the previous outputs of the system under test. This differs from static approaches (*batch testing*) which are more affected by the state explosion problem [2].

Overview The article is structured as follows. Section 2 is an overall description of the language of features and introduces the notion of behavioral pattern, which is an essential component of this description. In sections 3 and 4, we successively illustrate our method through an example in the fields of telecommunication systems. Section 5 concludes on this contribution.

2   Feature Construction in a Synchronous Context

We describe in this section the synchronous approach we propose to construct features. We then present the various specifications that define a feature. The operations of composition and integration are detailed in section 3.

2.1   Overview of the synchronous approach

Reactive programs are applications that continuously interact with their environment: to each input provided by the environment, the application reacts by emitting an output. Synchronous programs [3] are a sub-class of reactive programs which satisfy the *synchrony hypothesis*: every reaction of a synchronous program is instantaneous. Synchronous programs have cyclic behaviours: at each tick of a global clock (also called instant of time), all inputs are read and processed simultaneously, and all outputs are emitted.

The communication mode which comes with the synchronous approach is based on instantaneous broadcasting: all recipients of a message get it simultaneously and instantaneously. However, it is possible to weaken this synchrony in order to describe other modes of communication, such as message-passing [5].

The synchronous approach helps avoid the combinatorial explosion problem, which makes approaches based on parallel and communicating processes intractable in some cases. Indeed, all parallel components of a synchronous system react simultaneously and, thus, their executions are not intertwined [3].

Several case studies have shown that the synchronous approach was adapted to the modeling of telecommunication features [17,7], and more generally, to the modeling of some parts of telecommunication systems [10,12]. Indeed, such systems are reactive, since they shall react to any request from the users as fast as possible. Some of their components can even be viewed as synchronous [4,1]. At the level of abstraction we chose to favor, this view appears to be well adapted.

2.2 Feature Construction

A feature is an extension of a system. This system serves as a common basis for the integration of all features. By analogy with the terminology used in the fields of telecommunications, we denote this system as the "basic service".

A feature is not a stand-alone component, since it relies on the basic service in order to form a complete service, deliverable to the user. As a consequence, the modeling of each feature shall refer to the modeling of the basic service. On the other hand, the features are to be modeled in an isolated way, independently from one another.

In order to obtain a modular model, we provide an integration operation: given the models of the basic service and of one feature, this operation automatically produces the resulting service. An operation of composition is also needed, in order to integrate several features in a single system. This operation is more closely related to the validation process. These two operations could have been gathered in a single one: the first feature is integrated into the basic service, the second feature into the resulting service, ... However, separating them allowed for a more structured architecture and led to a more modular model.

We present in this section the manner to describe the various aspects of a feature. The composition and integration operations will be defined later on.

Description of a feature A feature F is characterized by a triple (B,R,P) where:

- B is the feature's behaviour.

- R is a set of requirements imposed on F (the properties that the feature is expected to satisfy).
- P is a set of behavioral patterns that describe how F is used by its subscribers.

Behaviour The behaviour B of a feature is an executable specification that corresponds to a set of I/O automata. It describes the extensions and the modifications that the feature will impose on the basic service. Since the basic service is supposed to serve as a basis for the features, it is necessary that its own states and transitions are clear and explicit, so that the feature's designer is able to determine which parts of the basic service are redefined by the feature[2].

Requirements The requirements R are a set of properties that the feature is expected to satisfy. To our experience, interesting properties can have two typical forms:

- invariants, e.g. "such situation never occurs". Such properties regard the state of the model and guarantee that some states are not reachable. For instance, a call forward feature is expected to prevent its subscriber's phone to ring at all time (since the calls intended for it must be redirected). This is stated informally as "the phone of a call forward subscriber never rings".
- behaviorals, e.g. "such behaviour leads always to such situation". Such properties regard the evolution of the state of the model and guarantee that some paths lead always to some states. We can keep as example the call forward feature and state that "the user dialing the number of a call forward subscriber will be put in communication with the user designated by the subscriber as target of the forwarding".

Behavioral patterns The behavioral patterns P describe the *usages* of a feature. It shall indicate the various manners available to the end user to activate or to invoke a feature. For example, the Call Forward feature mentioned above can be activated, deactivated and parameterized. All three tasks require different sequences of actions.

The purpose of behavioral patterns is to make the validation process more effective by driving the test generation towards the production of relevant sequences of events.

---

[2] In the context on telecommunication features, the states and transitions of the basic service are used to define the feature's Points of Invocation and Points of Return (cf. figure 2).

Definition of Behavioral Patterns A *behavioral pattern* (P) is made of successive alternating instant conditions and interval conditions. The instant conditions should hold one after the other, while the interval conditions should continuously hold on the corresponding interval[3].

A behavioral pattern represents a set of sequences. Any sequence such that the instant conditions are successively satisfied, while the corresponding interval condition continuously holds, is an instance of the pattern. When all the instant conditions have been observed one after the other in the given order, provided that the corresponding interval conditions have been continuously satisfied, the pattern is said to be *completed*.

A behavioral pattern completion can be recognized by an input/output automaton. Each transition stands for an instant or interval condition, while the states describe the various levels of progress according to the pattern. Let us consider a pattern, A *[B]* C *[¬ C]* A, where A, C and A are the successive instant conditions that are to be observed, and B and $\neg C$ the successive interval conditions that should hold in the meantime. The completion of this pattern can be recognized by the automaton depicted on figure 1, where I is the initial state and F is the final state, reached when the pattern is completed.

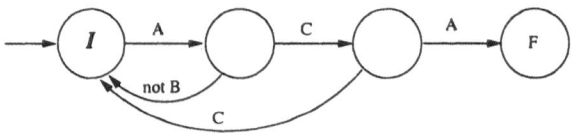

Fig. 1. Recognition automaton for A *[B]* C *[¬ C]* A

3   Modeling the Telephone System

This section presents our work regarding the modeling of telecommunication features. It is the result from several case studies. The modeling activity consists in extracting from a specification, more or less detailed and more or less formal, the relevant information that will allow its validation. In order to give a better credibility to our work, we chose to comply to the ITU-T recommendations, which we briefly describe below.

---

[3] We could have chosen to constrain intervals by stating what we don't want to observe. This choice does not affect the expressiveness of the language.

3.1   Context

According to the ITU-T[4][9], a feature should be viewed as an alternative to some parts of the basic service, usually called POTS (Plain Old Telephone Service).

A feature can take the control of a call in a given state of the POTS, denoted by POI (Point of Invocation). It gives back the control to the POTS in a state identified by POR (Point of Return). The figure 2 indicates how a feature can be related to the POTS.

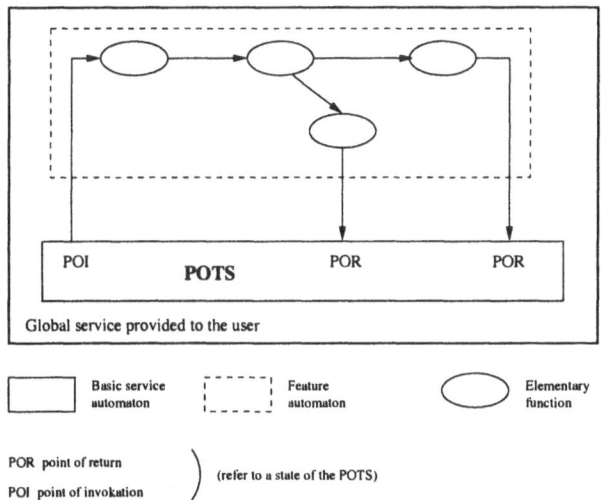

Fig. 2. Relation of a feature to the POTS in the Global Functional Plane

In the ITU-T recommendations, the processing of this basic service is described by means of two automata, each of which takes care of the interactions with one user. The ITU-T defines one Originating Basic Call Model (O-BCM) to describe the service viewed by the calling user, and one Terminating Basic Call Model (T-BCM) to deal with the called user's view. The automata are assumed to communicate via message-passing.

As a consequence, the global service provided by the telecommunication system is distributed and is represented by several automata. Each automaton has only a local influence within the system.

This view is close to reality, where two users involved in the same communication can be in distant places. We tried to keep our approach as close as possible to this view which states a principle of local control: there is no single entity to control a whole communication.

---

[4] Telecommunication Standardization sector of the International Telecommunication Union.

3.2 Overview of the Model

The model includes the telecommunication system and all the services provided by this system. Since our ultimate goal is the functional validation of the features, we can abstract any component that does not participate in the functioning of the services and features.

The model interacts with its environment, which is made of the set of telephones through which the final users can access the system. The model is assumed to react to any solicitation from is environment, i.e., it shall evolve quicker than the latter.

The outputs of the environment describe the actions performed by the users on the telephones (e.g. Go off the hook, Dial a number, ...), while the reactions of the model are the various tones that the telephones shall emit (e.g. Ringing tone, Busy tone, ...).

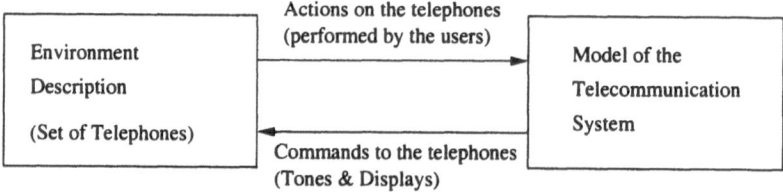

Fig. 3. Interactions between the model and its environment

Core of the model With each user outside the model is associated a *logical telephone* (LT) within the model. An LT is a behavioral specification of POTS and the features, as viewed by one particular user. An LT includes information about the state of the user's phone: what is its current output, who is its user's party, ... It also includes data about the features: which ones have been subscribed to by the user, and what are the parameters associated with each feature.

This notion of logical telephone is essential in order to express the principle of local control stated above (cf. section 3.1).

Communication between LTs LTs are synchronous and reactive. They all react to their inputs at the same pace. LTs need from time to time to communicate with one another. Indeed, two LTs can be momentarily involved in the same conversation and require to be synchronized. For instance, when a user dials a number, the reaction of the system depends on whether the called phone is busy or not. If it is not, it shall begin to ring, while the calling user shall hear a ring back tone.

Following the level of detail in the initial specification, the communications within the model can vary. For instance, in one of our case studies [7], the initial specification mentions no explicit communication within the system, while in an other one, it is clearly stated how informations is transmitted from one LT to another. The difference comes from the range of a message: in the first case, messages were to be provided at the same time to all LTs, while in the second, only one LT was the recipient of the message, and it was up to it to propagate this information to others.

The instantaneous broadcast communication of the synchronous approach is adapted to the first kind of architecture of the model. In that case, communications within the model are performed by making the states of LTs a set of global variables, so that any LT can consult the state of any other LT and react instantaneously in consequence.

In the second case, and whenever a message coming from the environment can lead to the generation of internal messages within the model, it appeared impossible to use the same kind of communication due to the problem of *causality*. Indeed, if communications are instantaneous, one LT could get a message from the environment, produce an internal message to ask another LT about its state, and get the answer from this LT, all at the very same instant. In other words, the LT is getting two messages at the same instant, one being the consequence of the other.

In order to avoid such conflicting situations, one simple solution consists to assign a non-null duration to any message. To guarantee the reactivity of the model (which is supposed to react quicker than its environment), we have to make sure that the environment cannot be able to produce new inputs before the end of the model's reaction. This can be ensured either statically, by computing a maximum reacting time of the model, or dynamically, by forbidding any action of the environment as long as the model does not explicitly indicate the end of its reaction.

3.3   Incremental Feature Composition and Integration

The operations of composition and integration are carried out in a homogeneous manner on all elements of the triple (B,R,P) describing the feature (cf. section 2.2).

Behaviour composition and integration The behavioural part of a feature can be composed of several automata. Each of them describes one generic behaviour that the feature can induce on one LT. In a general manner, one can distinguish the behaviours corresponding to the subscriber of the feature and the ones that correspond to the other users involved in the feature's operation.

Integration The behaviour that results from the integration of a feature $F = (B, R, P)$ into POTS is denoted by $\frac{B}{POTS}$. It consists in combining the automata of the basic service and of the feature. This combination is performed in the following way:

- The basis is the automaton A of the basic service.
- For each state of the automaton B (representing the feature), if the state is not in A (i.e., is not a POI nor a POR), add the state to the set of states in A
- For each transition t in B, let s be the originating state. If a transition from s with the same label than t exists, suppress this transition and replace it by t. Otherwise, add the new transition.

This operation implements a priority relation between the basic service and the features. Features are said to have priority over the basic service.

Composition The behaviour of the compound feature resulting from the composition of $F1 = (B1, R1, P1)$ and $F2 = (B2, R2, P2)$ is denoted by $B1 \oplus B2$. There are two solutions to perform this composition:

- Reuse the notion of priority introduced by the integration operation and construct the compound feature's behaviour as the integration of B2 to B1 (composition based on superimposition). Similarly to the integration operation, transitions in B2 may overwrite or complement the transitions in B1.
- Consider all features without priority (composition based on juxtaposition). In that case, the automata of the features are considered as being concurrent and transitions in B1 and in B2 are synchronized: if a transition is fireable in B1 and another in B2 (i.e., both automata react to the same input), both transitions are taken. If these transitions produce outputs, both outputs are emitted. In that case, the reaction of the LT can be inconsistent.

We preferred the first solution, as it allows to keep a consistent behavior of the LT. Note however that the second solution eases the interaction detection, since an inconsistent reaction can be seen as an interaction.

Modularity of the Composition/Integration In order to keep the model modular, the implementation of the model does not combine the automata of the features and service. Actually, the automata are run in parallel. The operation is as follows:

- Both the basic service automaton and automata B_i are fed with the same input, at the same time.
- All automata are executed in parallel, and their results are instantaneously available. Each automaton proposes a reaction or chooses to ignore irrelevant inputs.
- Given the priority order (among the features, and between the features and the basic service), an automaton reaction is selected. The automaton of highest priority to propose a reaction is chosen. The output of the LT is the one proposed by this reaction. Other automata are warned that their reactions have been rejected.

In order to take into account the rejection of its proposal, a feature automaton shall distinguish the transition function and the next state computation function. This way, the proposal of the feature is obtained by applying the transition function without modifying the state of the automaton. The state evolution is only performed after the selection of one proposal: if the feature is selected, its next state is computed, otherwise the state does not evolve and the transition proposal is ignored.

Feature Requirements Integration and Composition

Requirements Integration The properties of the basic service can be altered when a feature is integrated. This alteration corresponds to the fact that the feature is a modification of the basic service; in this context, it is quite natural that this modification has some consequences on the properties of the basic service.

This alteration consists in the suppression of some or all the basic service properties. It is assumed that the requirements of the basic service, as well as of any feature, are structured so that it can be seen as a conjunction of requirements or as a unique property. Suppressing some requirements comes down to eliminate some terms in the conjunction. We denote by $R_{POTS|F_i}$ the remaining properties resulting from the integration of the feature F_i to the basic service (POTS).

The properties of the feature are not modified by its integration. The properties of the service resulting from the integration of a feature $F = (B, R, P)$ to the POTS are therefore:

$$R_{POTS|F_i} \wedge R$$

Requirements Composition When composing two features $F1 = (B1, R1, P1)$ and $F2 = (B2, R2, P2)$, their requirements $R1$ and $R2$ are concurrent, since they were developed in an independent manner. The composition of requirements can't be operated on its own. This operation requires to first proceed to the integration of each feature to the POTS independently, in order to compute $R_{POTS|F1}$ and $R_{POTS|F2}$. As a consequence, the requirements of the compound feature is made of $R1$ and $R2$, taken separately from one another, and differs from the simple conjunction of requirements $R1 \wedge R2$.

The requirements resulting from the integration of $F1 \oplus F2$ to POTS are:

$$R1 \wedge R_{POTS|F1} \wedge R2 \wedge R_{POTS|F2}$$

Feature Patterns Integration and Composition In order to unify the integration and composition operations, we defined a language based on behavioral patterns. This language allows to combine patterns thanks to the following operators.

1. Priority operator. This operator is binary, commutative and associative. It is denoted "<>".

    Using a compound behavioral pattern $P_1 <> P_2$, the tester has a means to focus the guiding on one of the two patterns. The pattern that is able to start before the other gets a temporary priority over the other. The priority applies only up to the completion of this pattern. Another attempt to execute $P_1 <> P_2$ in the same test sequence can lead to another priority. An analogy can be found with the principle of critical section.

2. Concurrency operator. This operator is binary, commutative and associative. It is denoted "||".

    The concurrency operator is offered for expressing different usages which can take place concurrently. The progression of the patterns can be interleaved or simultaneous. Considering a compound behavioral pattern $P_1 || P_2$ instead of a simple pattern adds two steps to the regular guiding process, the first one as a preamble to the generation, the second one as a postamble:

    - Prior to the input generation, a pattern is chosen in a random manner. The regular generation is then conducted according to the selected pattern.
    - Once the input data has been generated, all patterns should be updated. Indeed, an input belonging to the progression category of the selected pattern may lead another pattern to regress or progress, which should be taken into account.

Feature Patterns Integration Strictly speaking, the integration of a feature $F = (B, R, P)$ to the POTS does not affect the patterns, since no pattern is associated with the basic service. However, there is still a combination to operate among the various patterns of the feature, in order to perform the validation of the isolated feature in an effective way.

This combination consists in applying the priority operator to the set of patterns. Each pattern describes a usage of the feature; combining them with the priority operator allows to provide a guiding following one usage at a time. Given $P = P1, ..., Pn$, the combination of patterns to be used to guide the test generation is $P1 <> ... <> Pn$.

Feature Patterns Composition To test a composition of features, the concurrency operation provides an interesting means of guiding the test generation. Indeed, all features are meant to be independently used, and to be accessed concurrently by various users. It is therefore useful to have a means to express this concurrency.

To effectively test a composition of n features $F_i = (Bi, Ri, Pi)$, it is recommended to use $P1 || ... || Pn$ as a guide for the test generation.

4   Validating Telephone Features

This section describes the process which was used to validate telecommunication features; it is however in no way dedicated to that specific kind of
features, and could be easily adapted to any other kind, provided that they
can be specified as described above.

4.1   Incremental Validation Principle

Roughly, the validation of a reactive system involve three specifications:

- M - the model of the system, which shall be executable and reactive.
  This model is constructed from the description of the basic service and
  of each feature provided by the system.
- R - the requirements, i.e. the properties that the users can expect from
  the system. This specification relies on the description of the features.
- E - the specification of the environment of the model.

The validation process consists in ensuring that M satisfies R, provided
that E is satisfied. The specification E of the environment is a model of
the real environment of the system. It provides some assertions about the
relations between the successive values of the inputs and outputs of the model
under test. The validation is based on a constrained random generation. It
produces test input data conforming to the constraints stated by E.

The validation follows an incremental process:

- The feature is first validated in isolation, as if it was the only feature
  available in addition to the basic service. It consists in applying the integration operation to that feature and validating the resulting model.
  This allows to ensure the internal correctness of the feature.
- The feature is then confronted to other features[5]. This is done by first
  applying the composition operation to that feature and to the other features, already provided by the model. Then, the same integration operation is performed. The testing of the resulting model allows to detect
  interactions among features.

4.2   Definition of a Feature Interaction

A widely accepted definition of an interaction is given by Combes [6]. This
definition is based on model satisfaction and states the following: given $POTS$
the model of the telecommunication system providing the basic service, $F_1...F_n$
n features, $B_1...B_n$ the behaviours of the features, $R1...R_n$ their requirements,

---

[5] Up to now, we deliberately limited ourselves to deal with pairs of features. We
discuss however in the concluding section of the possible extensions of this approach.

$\models$ a satisfaction relation between a model and a logical property, there exists an interaction among $F_1...F_n$ if and only if

$$\begin{cases} \forall i \in 1,..,n, POTS \oplus B_i \models R_i \\ POTS \oplus B_1... \oplus B_n \not\models \bigwedge_{i=1,..,n} R_i \end{cases}$$

This definition appears to us to have the following drawbacks:

- It does not distinguish among integration and composition of features, which we consider as necessary (cf. section 2.2).
- All properties are attached to features, none is provided regarding the basic service nor the underlying system.

In order to distinguish among integration and composition of features, we represent by $... \oplus ...$ the composition operation and by $\overset{...}{...}$ the integration operation. $B_1 \oplus ... \oplus B_n$ defines the behaviour of the compound feature resulting from the composition of features $F_1...F_n$. $\frac{B_1 \oplus ... \oplus B_n}{POTS}$ describes the integration of the compound feature to the basic service.

In addition to the feature requirements, we propose to take into account properties regarding the model without features. These properties can deal with the behaviour of the basic service or can define some consistency rules of the data within the model. These properties can possibly be invalidated by a feature: a feature being a modification of the basic feature, it is expected to modify to a certain extent the properties of the latter.

There are therefore two kinds of interactions in our approach:

- The first level is between one feature and the basic service and is observed when the feature invalidates the properties of the basic service. This kind of interactions is foreseen and does not correspond to an error. We denote R_{POTS} the set of properties of the basic service ; $R_{POTS|F_i}$ is the set of "remaining" properties, i.e. the properties of R_{POTS} that remain satisfied when the feature F_i is integrated to the basic service.
- The second level is between several features and occurs when features are composed. This kind of interactions is observed when one property of one feature is no longer satisfied or when one of the remaining properties for any feature is violated[6].

The interaction definition stated above can therefore be modified as follows:

$$\begin{cases} \frac{B_i}{POTS} \models R_i \wedge R_{POTS|F_i}, 1 \le i \le n & V_1 \\ \\ \frac{B_1 \oplus ... \oplus B_n}{POTS} \not\models \bigwedge_{i=1..n}(R_i \wedge R_{POTS|F_i}) & V_2 \end{cases}$$

V_1 corresponds to the validation of each feature in isolation, while V_2 corresponds to the confrontation of features.

---

[6] Notice that $R_{POTS|F_1} = R_{POTS|F_2} = R_{POTS}$ does not necessarily imply $R_{POTS|F_1 \oplus F_2} = R_{POTS}$.

4.3   Guiding of the Testing Process by Behavioral Patterns

For complex systems, reasonable behaviours of the environment may reduce to a small part of all possible ones with respect to the environment constraints (cf. section 4.1). Some interesting functions of a system may therefore be hardly tested since their execution may require sequences of actions which are too long and complex to be randomly frequent. To fix this drawback, one may wish to generate inputs in a more relevant way with respect to the most interesting sequences to be produced. To that aim, we propose to guide the generation with behavioral patterns. We define hereafter more precisely the notion of behavioral pattern, which has been introduced in section 2.2. We then describe more precisely the pattern-based guiding.

Definitions The environment is defined with three sets of boolean variables:

- O is the set of the system output variables (that is, the input variables of the environment),
- I is the set of the system input variables (that is, the output variables of the environment),
- L is the set of local variables that define the state of the environment.

In the following, for any set X of boolean variables, V_X denotes the set of values of the variables in X. $x \in V_X$ is an assignment of values to all variables in X.

Definition 1. We define the set of expressions Π_V on a set of variables V by:

- false $\in \Pi_V$, true $\in \Pi_V$
- $\forall p, p' \in V, p \in \Pi_V, \neg p \in \Pi_V, p \vee p' \in \Pi_V, p \wedge p' \in \Pi_V$

Definition 2. A *simple predicate* (SP) is built as the conjunction of an expression in $\Pi_{L \cup O}$ (named the state condition) at the previous instant[7] and an expression in Π_I (called the input condition) at the present instant:

$$SP = pre\ p_{L \cup O} \wedge p_I,\ with\ p_{L \cup O} \in \Pi_{L \cup O}\ and\ p_I \in \Pi_I$$

Definition 3. An *instant condition* (IC) is built using conjunctions, disjunctions and negations of simple predicates. Its syntax is defined as:
$$IC ::= SIC \mid SIC \wedge IC$$
$$SIC ::= NSP \mid NSP \vee SIC$$
$$NSP ::= SP \mid \neg\ SP$$

An instant condition allows the tester to express the fact that different situations can be expected at the same instant depending possibly on the responses

---

[7] *pre x* designates the value of x at the previous instant.

of the system under test. For example, the instant condition below means that if the output was $RingingTone_A$ at the previous instant, Off_A is expected to be the input at the current instant:

$$pre\ RingingTone_A \wedge Off_A$$

Definition 4. An interval condition (LC) is defined as a simple predicate.

Definition 5. A *behavioral pattern* (BP) is made of alternating instant conditions and interval conditions:
$$BP ::= [LC]\ IC\ |\ [LC]\ IC\ BP$$

Note that while in practice the first interval condition is often set to *true*, it is in no way mandatory. X[Y]Z means "X is expected to hold, then Z, while Y holds continuously in the meantime".

For instance, the following behavioral pattern expresses that A is expected to call B:

$$Idle_A \wedge Off_A\ [\neg On_A]\ Dial_A(B)$$

$Idle_A$ is a state predicate that indicates that phone of user A is on the hook and not ringing. Off_A, On_A and $Dial_A(B)$ describe inputs from the environment stating respectively that A goes off the hook, A goes on the hook and A dials the number associated to user B. The corresponding sequences are related to various situations: B may or may not accept the call, or may be busy at the time of the call arrival, ...

Characterizing a behavioral pattern The completion of a behavioral pattern can be interpreted as a regular expression over the simple predicates that are involved. Given a behavioral pattern BP such that

$$BP = [inter_0]cond_0[inter_1]cond_1 \ldots cond_{n-1}[inter_n]cond_n,$$
the regular expression describing its completion is:
$$(inter_0 \wedge \neg cond_0)^*.cond_0.(inter_1 \wedge \neg cond_1)^*.cond_1.$$
$$\ldots cond_{n-1}(inter_n \wedge \neg cond_n)^*.cond_n$$

For instance, given A, B, C three booleans inputs and a behavioral pattern $A\ [B]\ C\ [\neg\ C]\ A$, the input sequence AACBA is matching, while ACCA is not.

Behavioral patterns can therefore be characterized as well in a past temporal logic formalism: expressivenesses of regular expressions and of past temporal logic have been shown to be equivalent [13]. As a result, patterns can be used as a basis to build behavioral properties (cf. section 2.2).

The construction of the past temporal formula corresponding to a behavioral pattern is performed in an incremental way, by characterizing each instant condition with respect to the previous one. For instance, the pattern above, $A\ [B]\ C\ [\neg\ C]\ A$, is described as follows, where **after, between, always_since** are widely-used past temporal operators:

$$P_1 = A$$

$$P_2 = \text{C and } \texttt{between}(P_1,\ P_2) \text{ and } \texttt{always\_since}(\text{B},P_1)$$
$$P_3 = \text{A and } \texttt{between}(P_2,\ P_3) \text{ and } \texttt{always\_since}(\text{not } \text{C},P_2)$$

P_3 being the predicate indicating the completion of the pattern.

Pattern-based Guiding A behavioral pattern can serve to guide the input data generation process as follows: among the data satisfying the environment constraints, those meeting the conditions are preferred over the others. However, since the data are randomly selected, the latter ones have still a chance to occur.

While the environment specification sets physical or model-related constraints that cannot be avoided, the behavioral patterns are weaker constraints aiming at driving loosely the generation.

In practice, the pattern-guided generation consists in keeping up to date a progression index on the considered pattern. At each step, all inputs that satisfy the instant condition designated by the index see their occurrence probability increase, while the inputs that satisfy neither the instant condition nor the interval condition designated by the index have their probability reduced. The progression index indicates what prefix of the pattern has been satisfied so far. To any value this variable can take corresponds a pair of predicates {*inter, cond*} which describes the next-to-appear instant condition (*cond*) and the predicate that should continuously hold in the meantime (interval condition, *inter*).

For each state of the environment and for a given value of the progression index, the set of possible inputs (i.e., inputs that satisfy the environment constraints) is divided in three categories:

- $\mathcal{P}$ (*progression*) includes all inputs that make the guiding process go forward,
- $\mathcal{N}$ (*neutral*) gathers all inputs that do not affect the guiding process,
- $\mathcal{R}$ (*regression*) groups the inputs that lead to the process abortion.

A probability is assigned to each category so that input vectors in the first one would be favored over input vectors in the second category, which, themselves, would be preferred to input vectors from the third category. These probabilities are determined with respect to the cardinality of each partition and to given weights associated with them : $w_{\mathcal{P}}$, $w_{\mathcal{R}}$ and $w_{\mathcal{N}}$. A partition is said to be of higher priority than an other if its weight is greater.

The input selection is a two-step process. First, a category is selected according to the determined probabilities. Each category c in $\mathcal{C}=\{S_{\mathcal{P}},\ S_{\mathcal{R}},\ S_{\mathcal{N}}\}$ has a probability p_c of being selected:

$$p_c = \frac{w_c * card(c)}{\sum_{j \in \mathcal{C}} w_j * card(j)}$$

Then, an input vector is chosen in an equally probable manner from the selected category. As a result, the probability for any input vector i in c to be chosen is $p_{i,c}$:

$$p_{i,c} = \frac{1}{card(c)} * p_c = \frac{w_c}{\sum_{j \in \mathcal{C}} w_j * card(j)}$$

4.4 Application

We applied this approach to feature interaction detection by participating to the First Feature Interaction Detection Contest [7]. The approach turned out to be well adapted to the problem, since it led us to detect the most interactions and win the contest. In particular, the use of behavioral patterns contributed highly to our success, since they appeared necessary in the detection of 2/3 of the interactions. The language based on patterns that we developed to guide the test generation allowed a significant relief for the human tester in this case study: there were 12 features in the contest, and features were to be composed and validated by pairs, thus leading to the analysis of 78 configurations. Combining manually the behavioral patterns would have been a tedious work, which we avoid thanks to the concurrency and priority operators on patterns.

5 Conclusion

The construction of feature-based systems is an emerging research theme. We describe in this article an experiment and a proposition for a unified and incremental tackling of feature modeling and validation. The approach we propose is based on a highly modular modeling which allows each feature to remain independent. Yet, the relation between one feature and the remaining of the system is clearly defined. The validation process takes advantage of a well-adapted testing technique based on behavioral patterns. A language is defined to combine patterns in order to test effectively each combination of features and to unify the modeling and validation approaches. This method has been applied to feature interaction detection in the fields of telecommunication systems and turned out to be well adapted to the problem.

6 Acknowledgements

The author would like to thank Farid Ouabdesselam for his helpful comments and advice of improvements on the previous drafts of this paper.

References

1. P.K. Au and J.M. Atlee. Evaluation of a state-based model of feature interactions. In *Feature Interactions in Telecommunications Systems IV*, pages 153–167. IOS Press, 1997.
2. A. Belinfante, J. Feenstra, R. G. de Vries, J. Tretmans, N. Goga, L. Feijs, S. Mauw, and L. Heerink. Formal test automation: a simple experiment. In *12th International Workshop on Testing of Communicating Systems*, pages 179–196. Kluwer Academic Publishers, 1999.

3. A. Benveniste and G. Berry. The synchronous approach to reactive and real-time systems. *Proceedings of the IEEE*, 79(9):1270–1282, 1991.

4. J. Blom, R. Bol, and L. Kempe. Automatic detection of feature interactions in temporal logic. In *Feature Interactions in Telecommunications Systems III*, pages 1–20. IOS Press, 1995.

5. F. Boniol and M. Adelantado. Programming distributed reactive systems: a strong and weak synchronous coupling. In *7th International Workshop on Distributed Algorithms*, pages 294–308. LNCS, september 1993.

6. P. Combes and S. Pickin. Formalization of a user view of network and services for feature interaction detection. In *Feature Interactions in Telecommunications Systems*, pages 120–135. IOS Press, 1994.

7. L. du Bousquet, F. Ouabdesselam, J.-L. Richier, and N. Zuanon. Feature interaction detection using synchronous approach and testing. *Computer Networks and ISDN Systems*, 32(4):419–431, april 2000.

8. R.J. Hall. Feature combination and interaction detection via foreground/background models. In *Feature Interactions in Telecommunications Systems V*, pages 232–246. IOS Press, 1998.

9. ITU-T. Principles of intelligent network architecture. Recommandation Q.1201, 1993.

10. L.J. Jagadeesan, A. Porter, C. Puchol, J.C. Ramming, and L. Votta. Specification-based testing of reactive software: tools and experiments. In *19th International Conference on Software Engineering*, pages 525–535. ACM, 1997.

11. B. Kelly, M. Crowther, J. Kling, R. Masson, and J. DeLapeyre. Service validation and testing. In *Feature Interactions in Telecommunications Systems III*, pages 173–184. IOS Press, 1995.

12. G. Murakami and R. Sethi. Terminal call processing in Esterel. In *World Computer Congress*, Madrid, Spain, 1992. IFIP.

13. D. Pilaud and N. Halbwachs. From a synchronous declarative language to a temporal logic dealing with multiform time. In *Symposium on Formal Techniques in Real Time and Fault Tolerant Systems*. Springer Verlag, 1988.

14. M. Plath and M. D. Ryan. Plug-and-play features. In *Feature Interactions in Telecommunications Systems V*, pages 150–164. IOS Press, 1998.

15. M. Plath and M. D. Ryan. Feature integration using a feature construct. *Science of Computer Programming*, 2001. To appear.

16. H. Velthuijsen. Issues of non-monotonicity in feature-interaction detection. In *Feature Interactions in Telecommunications Systems III*, pages 31–42. IOS Press, 1995.

17. N. Zuanon. Test de spécifications de services de télécommunication. Phd thesis, Université Joseph Fourier, Grenoble, France, june 2000.

Author index